MW01641068

THE ART AND MAKING OF
STAR WARS®
THE OLD REPUBLIC

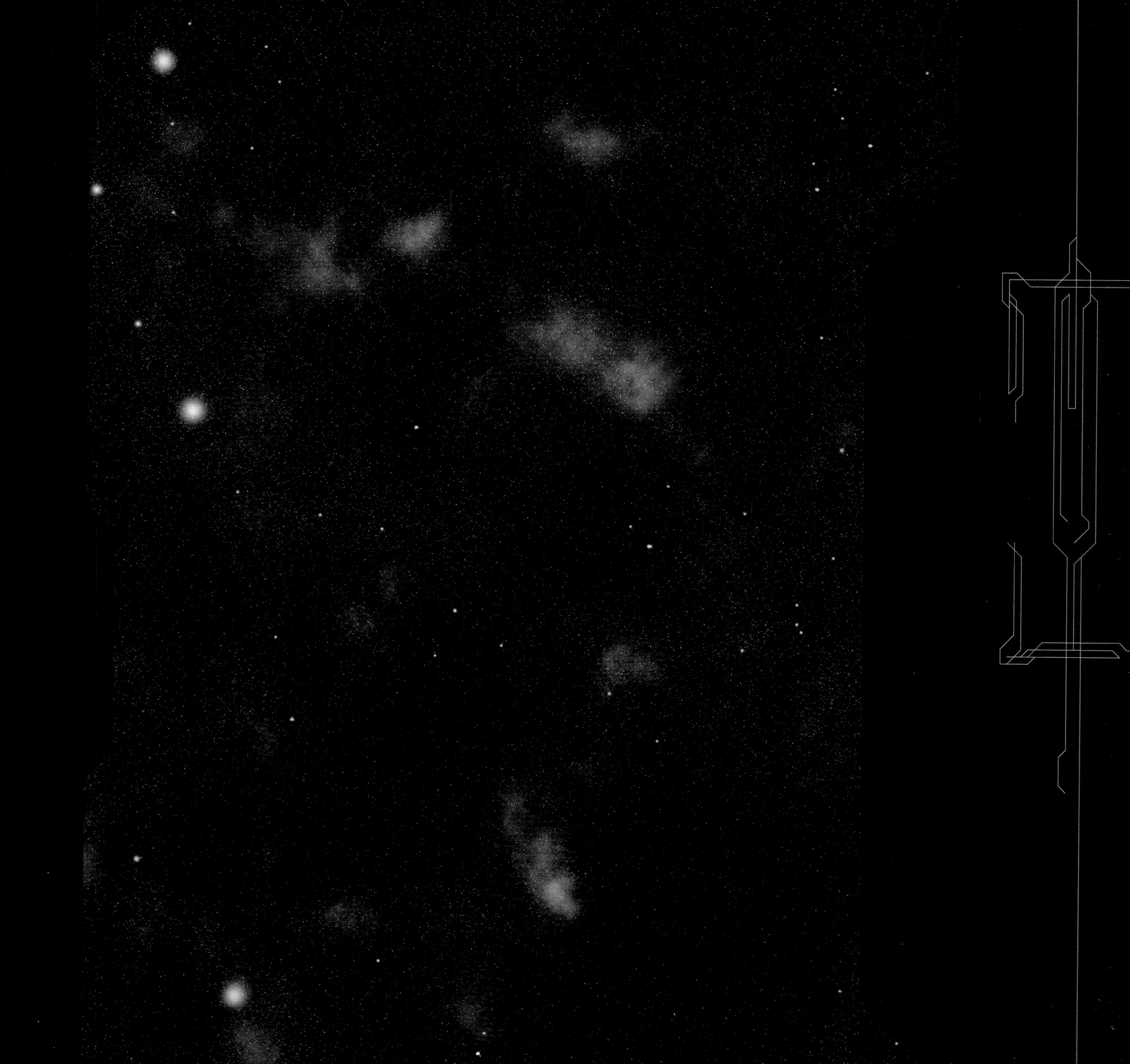

THE ART AND MAKING OF STAR WARS THE OLD REPUBLIC

BY FRANK PARISI AND DANIEL ERICKSON
FOREWORD BY PENNY ARCADE

CHRONICLE BOOKS
SAN FRANCISCO

Library of Congress
Cataloging-in-Publication Data:

Parisi, Frank.

The art and making of Star Wars :
the Old Republic / by Frank Parisi and
Daniel Erickson ; foreword by Penny Arcade.

p. cm.

ISBN 978-0-8118-7500-4 (hardcover)

1. Star Wars: The Roleplaying Game (Game)
2. Star Wars, the old Republic.
I. Erickson, Daniel.
II. Title.

GV1469.62.S7P37 2011

793.93—dc22

2011005942

Manufactured in China.

Designed by Public

10 9 8 7 6 5 4 3 2 1

Chronicle Books LLC
680 Second Street
San Francisco
California 94107
www.chroniclebooks.com

www.starwars.com

PAGE 1:
Early exploration of the Sith Academy
by Clint Young.

PAGE 2–3:
Look and feel piece created by Ryan Dening
to set visual target for the space game.

PAGE 7:
The Alderaan Warzone would eventually
have more ships and less large structures
than this first piece by Ryan Dening.

PAGE 8–9:
Clint Young's initial take on Tython was of
wild and unexplored forest.

For Curtis Cannell,
our friend and fellow lover of games.
You will be missed.

Clint Young's initial take on Tython was of wild and unexplored forest.

TABLE OF CONTENTS

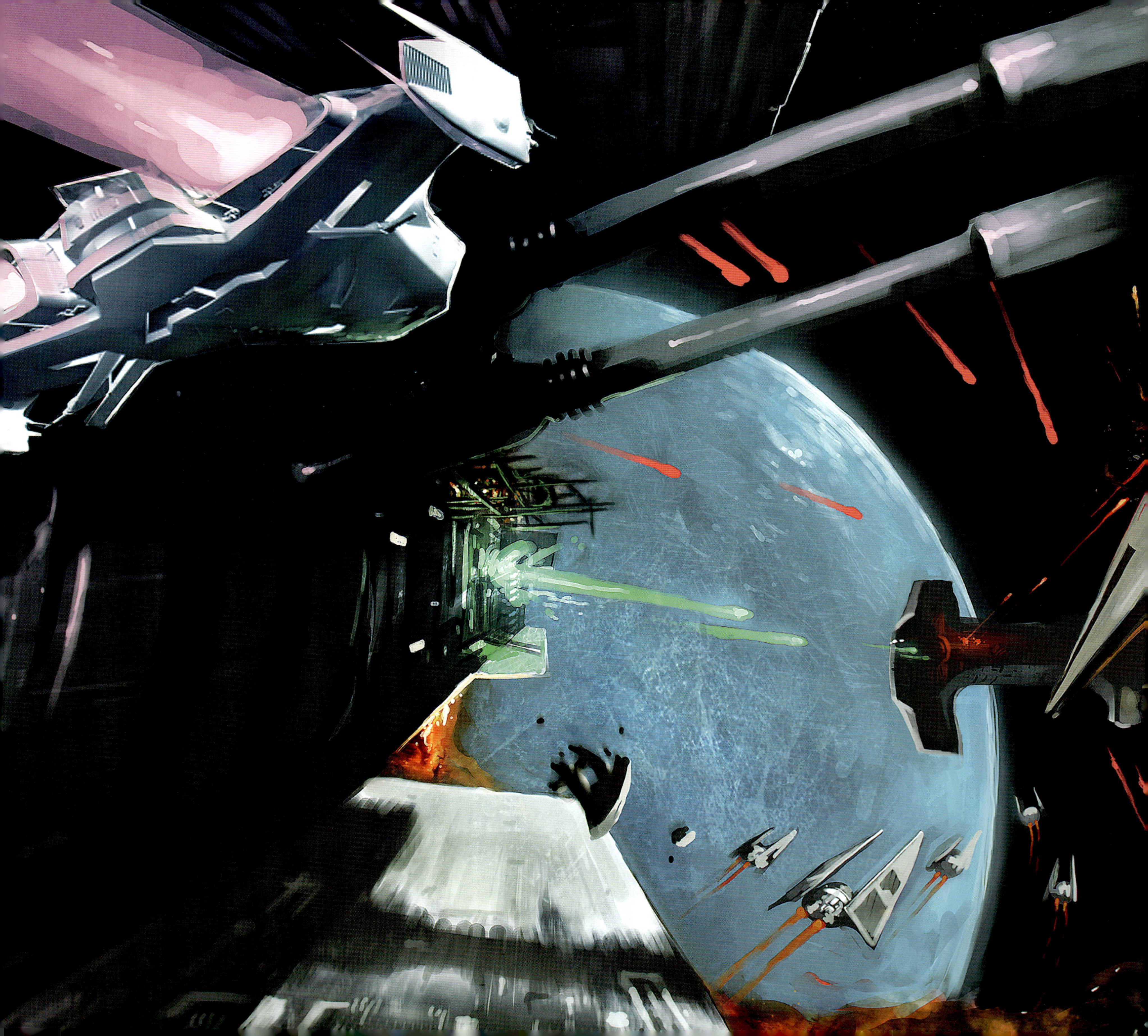

A Casper Konefal concept of one of the many space battle missions in the game.

FOREWORD

PENNY ARCADE

When I was asked if I would be interested in writing a foreword for *The Art and Making of Star Wars: The Old Republic,* I assumed there was a mistake. It's not uncommon for people to get Tycho and me mixed up. I can't count the number of times someone has come up to us at a convention and told Tycho what a fan they are of his artwork, or complimented me on my incredible vocabulary. The difference is that I am always quick to correct them, while Tycho simply smiles and accepts their praise. He is always so humble when taking credit for work that is not his.

This, however, was not a misunderstanding. They really wanted me and not him. After the shock wore off, I got right to work. The first thing I did was make some posters to hang around the office. Nothing fancy, just announcements that I had been tapped to write the foreword to a book and Tycho had not. Then some business cards with the job title *Head Writer in Charge of Writing.* When Tycho tore up the one I handed him, I decided to just leave a box on his desk. With the real work out of the way, I could concentrate on writing, and that's when the terror hit.

I sat paralyzed at my computer struggling to chain even a handful of words together. I don't write things; I draw them. I honestly had no idea how I was going to fill my 750-word quota. I was lucky that the title of the book was ridiculously long. Just look at it: *The Art and Making of Star Wars: The Old Republic.* I mean, that's ten words right there. Some quick work with a calculator told me that all I had to do was write the title seventy-five times and I'd be golden.

Needless to say, my first draft was rejected.

With my deadline fast approaching, I went to work on a second draft with the intention of actually writing something. I thought this was a very noble goal and decided to reward myself with some time spent actually playing *Star Wars: The Old Republic,* the game on which this book, *The Art and Making of Star Wars: The Old Republic,* is based (see what I did there). The hours slipped into days, and before I knew what was happening, my bounty hunter Ghorus Zintaal (most people just call him Gore) had an assortment of devastating powers and weapons. But I was no closer to actually writing anything. The fear of missing my deadline was overwhelming, so I decided to try to calm my nerves by playing a bit of my new favorite game, *Star Wars:* The Old Republic. Maybe I'd start a new character.

My Smuggler Dalia Chance was a real piece of work. She proved herself to be an outlaw with a heart of gold. She played both sides, but when she thought no one was looking, she always made the right decision, even if it meant not getting the most credits. Her epic journey was a tale worthy of the big screen. It would need to be a trilogy of course, and Kate Beckinsale would have to star. I was putting the finishing touches on some simple storyboards to accompany the screenplay I had written when I remembered the foreword I still needed to write.

By now the deadline had come and gone, and I was ashamed of myself. I spent hours thumbing through the pages of this book marveling at the incredible artwork. *Star Wars* is an inspiration for many artists, but this book is different. What these artists have created is an inspiration for *Star Wars.* The Old Republic might take place in the past, but this book represents the future of *Star Wars.*

Understanding the scope and power of what these artists have created led me to a period of introspection that I now refer to as "the black days." The conclusion I came to was that I am a terrible person—lazy, weak-willed, and ungrateful. The depression I fell into was like a deep, dark well. I felt cold inside and spent my days in bed under piles of blankets just sobbing quietly. After weeks of hardly eating, I decided that I needed to fight my demons rather than hide from them. So I started playing *Star Wars:* The Old Republic again.

My Jedi Ventus Kin had become a true Master of the force. His flirtation with the dark side mirrored my own internal struggles. In the end, his redemption proved to me that I could change as well. He showed me that . . . woah, that's 750 words! See you later suckers, I need to roll a Sith!

— Gabe

1

THE BEGINNING

"For over a thousand generations the Jedi Knights were the guardians of peace and justice in the Old Republic."

Obi-Wan Kenobi

Star Wars:
Episode IV *A New Hope*

"Deciding what project would be BioWare's first MMO was easy. We all love *Star Wars*. But when we got together with Lucasfilm again, the two companies decided to do something crazy," says Rich Vogel. The executive producer of *Star Wars*: The Old Republic is specifically referring to the decision to make the game a fully voiced, story-based RPG set in the ancient *Star Wars* galaxy in the spirit of BioWare's single-player titles. But it's a sentiment shared by everyone involved in making the electronic juggernaut. Let's not pull punches: The Old Republic is one of the biggest games ever created. Before dismissing this as hyperbole, consider: It's a full-fledged massively multiplayer online role-playing game (MMORPG) capable of supporting millions of players. Nineteen sprawling worlds with distinct inhabitants, histories, and technologies. Hundreds of pages of documents chronicling three hundred years of backstory. Professionally staged cinematics and fast-paced, choreographed combat. Dozens of companion characters that players get to know as real people, deeply and over time. Thousands of outfits, weaponry, abilities, and Force powers. Eight player classes, each with hundreds of hours' worth of story that twists and turns based on player choices. Over nine hours of music that twists and turns based on player choices. More than sixty novels' worth of voice-acted dialogue ("Forty man-years of writing," as lead writer Daniel Erickson puts it) that twists and . . . well, you get the idea. Oh, and on top of that, a space combat game.

And every bit of it looks, sounds, and feels like *Star Wars*.

Yep. *Crazy*.

PAGE 12–13:
Darths Revan and Malak from a timeline piece by Sperasoft.

The very first pieces of concept art by Arnie Jorgenensen were done in a strongly comic book 70s throwback style.

Accomplishing such an aesthetic, technological, and storytelling feat required a Herculean effort from more than three hundred designers, programmers, artists, testers, writers, animators, sound engineers, producers, managers, and composers at the top of their respective crafts. The goal: To build a teeming, breathing *Star Wars* galaxy in which people could not only live but also *create* their own personal *Star Wars* saga.

Welcome back to a galaxy far, far away. . . .

KNIGHTS OF THE OLD REPUBLIC

In 1999, game developer BioWare, of Edmonton, Alberta, Canada, had just come off the successful launch of its seminal fantasy role-playing video game, Baldur's Gate. California-based LucasArts, the game division of Lucasfilm, impressed with Baldur's nonlinear storytelling and flexible, choice-based character progression, approached BioWare about taking its game-play mechanics and injecting them into *Star Wars*. At the time, James Ohlen—the visionary lead designer of Baldur's Gate and eventual creative director of The Old Republic—was working on Neverwinter Nights, another high-fantasy game, and relished the opportunity to work with the unique genre meld *Star Wars* offered. "*Star Wars* is really fantasy in a sci-fi setting, and the fantasy is very subtle," says Ohlen. "The fact that *Star Wars* is differentiated from the rest of the fantasy out there would allow us to have a role-playing game that was quite a bit different from what we had done before."

Hands were shaken, papers signed, sleeves rolled. BioWare would handle the bulk of the game's development while LucasArts would spearhead a massive audio effort by providing sound effects and original music, as well as casting and recording the game's voluminous dialogue.

To give the designers free rein to craft a vast story without bumping against too much existing *Star Wars* canon, the game was set in the Old Republic era, four thousand years before the time of Luke Skywalker and Darth Vader. It was a huge gamble for LucasArts, which had yet to test the appeal of a game so far removed from the movies. "The challenge when we first started working with BioWare was, *Would the audience find appeal in a segment of the* Star Wars *universe that isn't reflected in any of the films?*" recalls Mary Bihr, LucasArts vice president of global publishing. "Maybe archetypes weren't enough. Maybe people wanted to play as Luke. Maybe they wanted to play Han Solo. Maybe they wanted Darth Vader in the game."

But the gamble paid off. Released in 2003, Knights of the Old Republic (KotOR) garnered critical praise for its sprawling story line, deep customization, robust voice-acting, and musical score, and won dozens of Game of the Year awards. According to Bihr, instrumental to the game's success was the way in which KotOR "surgically removed" archetypes and themes from *Star Wars* that resonated with fans and transplanted them into a new time frame fully intact. "That's mastery, when you think about it," Bihr says. "It requires a certain type of sensitivity and a deep, deep knowledge of *Star Wars*."

After KotOR's release, BioWare spent the ensuing years challenging and progressing Western notions of what RPGs were capable of, resulting in Jade Empire (2005), Mass Effect (2007), and Dragon Age: Origins (2009).

Original concept art of Darth Malak from Knights of the Old Republic by Mike Sass.

Though the team never thought they were doing a fantasy game, explorations like this one by Arnie Jorgensen allowed them to develop the ideas that would later be used for aliens.

A BRIEF MMORPG PRIMER

Massively multiplayer online role-playing games take place in enormous, virtual worlds, inside of which thousands of players—represented by customizable avatars—interact with one another. These environmental simulacrums, typically fantasy milieus, have the trappings of the real world. They evolve as players interact with them, and vice versa, and inhabitants and events persist whether anyone is playing or not. They are worlds of consequence, meaning that actions and choices, once made, irrevocably change the world. Designed to be social gaming experiences, MMORPGs allow players to trade virtual goods or services and to group together to engage in combat against other player characters (PCs) or nonplayer characters (NPCs) controlled by the game itself. As a result of their actions, characters accrue experience points and virtual currency, which can be allotted toward better abilities and more gear.

A barbarian rough conceived by Arnie Jorgensen during the brief pre-*Star Wars* era development of the game.

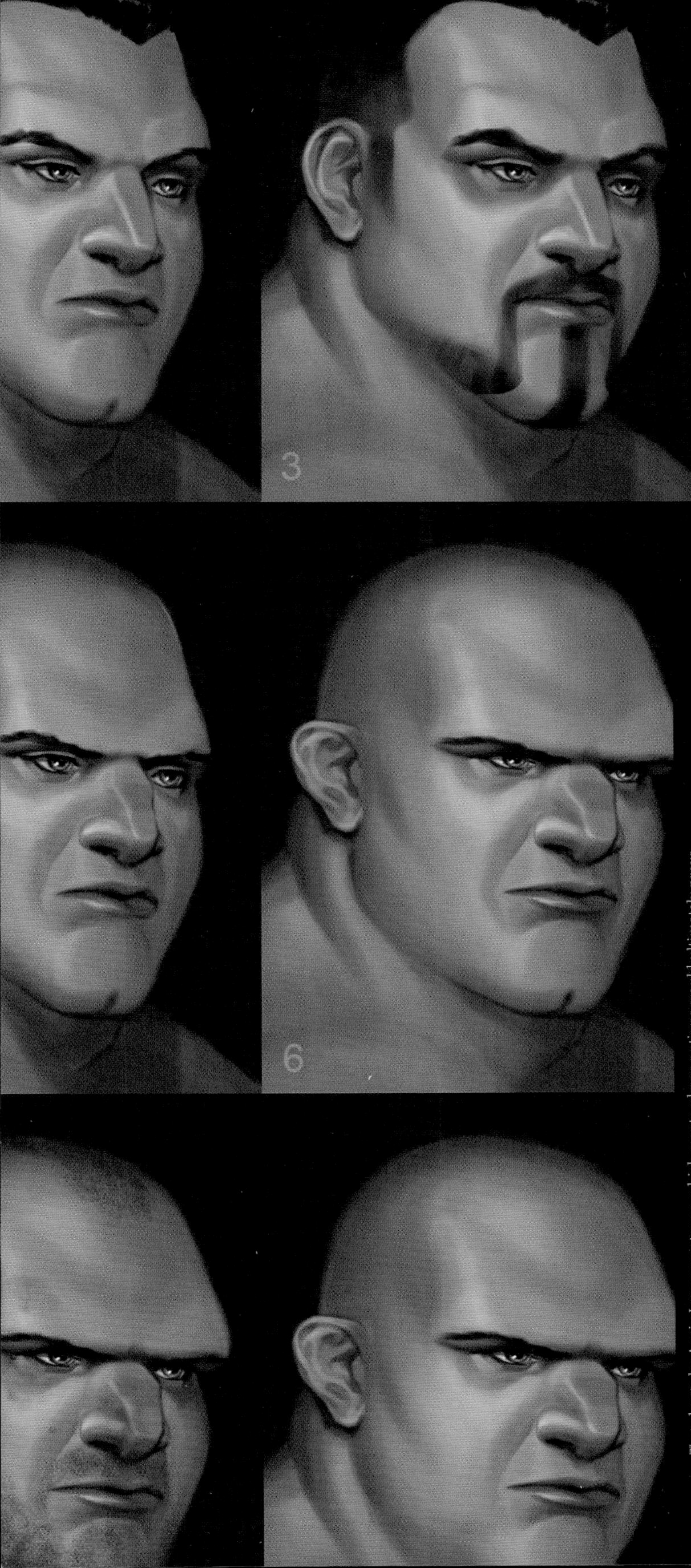

Test sheets by Arnie Jorgensen to see which create player options would ultimately prove the most impactful.

Female body explorations by Arnie Jorgensen for a far more exaggerated art style that was ultimately abandoned.

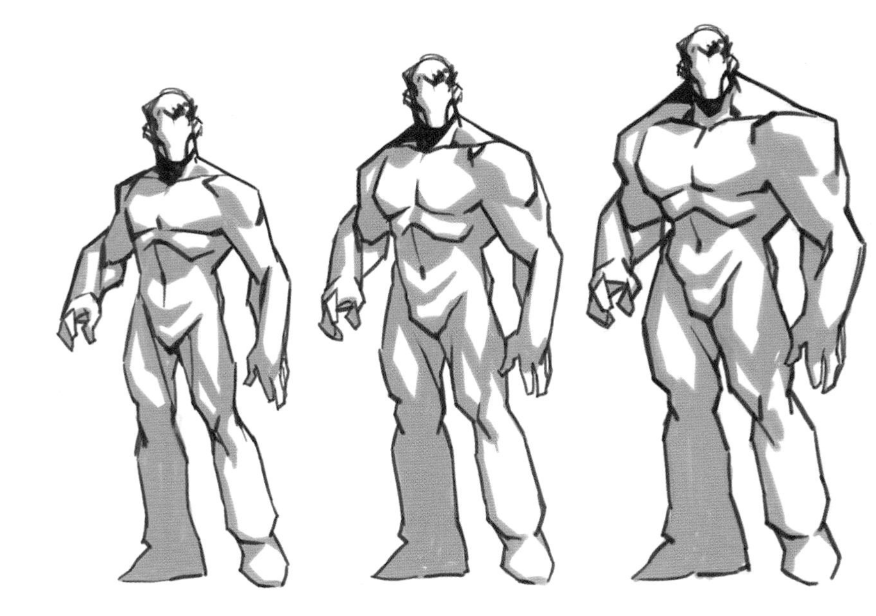

A discarded exploration by Arnie Jorgensen of far more extreme body types than those that would eventually be used.

MOVING INTO THE MMO SPACE

Even as far back as Neverwinter Nights, BioWare had considered moving into massively multiplayer online gaming. Although Neverwinter enabled players to effectively set up their own mini-MMOs and persistent worlds, it didn't match the scope that the company envisioned. "It did have its toolset and was multiplayer, and people would actually use the toolset to create small Neverwinter Nights MMOs," says Ohlen. "But nothing, obviously, to the scale of a real MMO."

In 2005 BioWare secured capital from the private equity fund Elevation Partners to establish a new studio dedicated to creating a full-fledged MMO. To do so, BioWare founders Ray Muzyka and Greg Zeschuk would need partners who could build a team specifically tailored for MMO development, the single most complicated game-creation challenge in the business. "There was no way BioWare was going to be able to do it on our own with the people we had, because no one knew the first thing about how to start building an MMO," says Ohlen.

Serendipitously, veteran MMO developers Rich Vogel and Gordon Walton approached Muzyka and Zeschuk at the Game Developers Conference in 2005 with the idea of building a BioWare-branded MMO. At the time, Walton and Vogel, based in Austin, Texas, who had worked on a number of MMOs including Ultima Online and *Star Wars:* Galaxies, were also looking to breathe new life into the genre. "Rich and I said, 'This isn't as much fun as it's supposed to be. We need to find someone who's serious about what we're doing here,'" remembers Walton. "World of Warcraft had come out, and it was very clear what the lessons were. You've got to put out something awesome. Not something half-baked."

Walton and Vogel agreed to build and lead what came to be BioWare Austin. An epicenter of MMO development, Austin had abundant talent readily available, and Walton and Vogel had their fingers on the pulse of the development community there. "We wanted to get a lot of talented, experienced MMO developers who had done it before, and Rich and Gordon were very hooked into the Austin scene and knew a lot of people who had delivered MMOs," says Zeschuk. Walton adds, "Austin was the right place for talent. We needed to be located here to make the next big online game."

In addition to the new talent recruited from the Austin area, Zeschuk and Muzyka began selecting key personnel from the Edmonton studio, including Daniel Erickson and Emmanuel Lusinchi, both of whom had just come off of senior design duties on Dragon Age: Origins, and James Ohlen. Erickson remembers the first creative meetings with Muzyka; Zeschuk; Ohlen; Lusinchi; Vogel; Walton; lead concept artist Arnie Jorgensen and art director Jeff Dobson, both of whom worked on *Star Wars*: Galaxies; and former technical director Bill Dalton: "Literally, the Austin studio started with us meeting in a tiny hotel room, hashing out the details of this huge, crazy project we were going to try and make."

Early ideas for the theme of the game included a "dark ages" time for the Republic as shown in this Jedi Knight by Arnie Jorgensen.

An early sketch of a Jedi by Arnie Jorgensen from the discarded "dark ages" theme.

RETURNING TO THE OLD REPUBLIC

Though BioWare explored a variety of ideas, creating an MMO follow-up to Knights of the Old Republic had huge appeal. "At the time, BioWare was known for KotOR," says Vogel. "Baldur's Gate and such were big games, but KotOR was the one that broke out, made people more aware of the BioWare name. We felt, *Why not look at that as an opportunity?*"

"We loved working in the *Star Wars* universe and working with LucasArts on Knights of the Old Republic," adds Muzyka. "We felt setting an MMO in that time frame would be something really special. The earlier prehistory of the movies is rich with lots of Jedi, lots of Sith, lots of unexplored territory."

From a development standpoint, revisiting *Star Wars* had practical benefits, too, given that creating an entire world as deep and complex as those in BioWare's games would have added years to production. "We thought it was important that we put everything we had into making the best story and the best game we could in an amount of time that would make it feasible to do a massively multiplayer online game," says Erickson. "We needed something developed, but with a lot of freedom." Most of all, it was a great fit creatively. "The most important thing is that whoever is sitting in your director's seat—in our case, James Ohlen—is very into the idea of what you're doing. James loves *Star Wars. The Empire Strikes Back* is like the cornerstone of his fantasy, and many of the characters from Knights of the Old Republic had come from the long-running pen-and-paper *Star Wars* games he played as a kid," Erickson says. "The man *loves Star Wars*."

"Knights of the Old Republic was very popular. People look fondly upon it," says Ohlen. "If you were to ask BioWare fans, 'Hey, BioWare is about to do an MMO, what MMO do you want them to do?' Most of our fans would say, 'Knights of the Old Republic MMO.'"

Meanwhile, LucasArts was planning to return to the massively multiplayer online space and exceed what it had accomplished with *Star Wars*: Galaxies.

"We were both looking to try and do one of these games," says Zeschuk. "Obviously, the fact that we were so successful together previously made it a lot of sense to reunite." Consequently, BioWare reached out to LucasArts in 2006 with the idea of collaborating again, this time on a rich, story-driven MMO set in the Old Republic era. "BioWare came in, pitched the game, and their proposal from the outset was strong," says Jake Neri, former producer at LucasArts. "The blend of *Star Wars* and BioWare is really powerful."

A *Star Wars* MMO with the narrative nuances of a BioWare RPG was too appealing a prospect for either company to pass up. "At the end of the day we're just big *Star Wars* fans," says Muzyka. "We're big RPG and MMO fans, so we're really passionate about this." LucasArts' Mary Bihr adds, "It's exciting working again with a team of master craftsmen in the art of storytelling and the ways in which story evolves in multiple directions. We've found a spiritual match in their ability to create worlds and characters that resonate with the audience."

MASSING THE TROOPS

The next step was to assemble the production team at LucasArts and the new BioWare studio in Austin. As with KotOR, BioWare would be responsible for the bulk of development: script writing, designing characters and environments, world building, animating, programming, and building all the technical systems. Later in the project, Electronic Arts, BioWare's parent company, would take over marketing, publishing, and external production duties as well, but in the early days of development these were all handled by LucasArts.

On the production side, LucasArts would also contribute its expertise in audio design to what would be the first fully voice-acted MMO. The team included director of audio Darragh O'Farrell (KotOR I and II, *Star Wars*: The Force Unleashed, the Monkey Island series), producer Orion Kellogg, and voice director William Beckman. Together they would cast, direct, record, and edit more than three hundred thousand lines of dialogue. For the soundtrack, music supervisor and staff composer Jesse Harlin would work with seven other composers on more than six hours of original cantina and orchestral *Star Wars* music.

"What we do is work with the producers at BioWare to make sure that our game is on target for what Lucasfilm expects," says Neri. Neri and LucasArts associate producer Tim Temmerman called on their experience working on *Star Wars*: Galaxies to help guide the game's development: coordinating and approving content, providing ideas and feedback, and ensuring that the game meet player expectations and Lucasfilm's approval. Working with Lucas Licensing, they also ensured that all continuity mesh with existing canon and that the game *feel* like *Star Wars*.

An early character exploration by Arnie Jorgensen done in a generic style because the *Star Wars* contract had not yet been signed.

"When you're writing interactive fiction, the work is game design as much as storytelling," says lead writer Daniel Erickson. What this means is that many of the normal rules of writing go out the window. First, there is no protagonist. The player could be playing as a man or a woman and see themselves as smart, funny, or mean. The idea of writing *Hamlet* starring a prince of Denmark who could be a shy but friendly gal with a forgiving nature requires a unique set of storytelling skills. The secret, says Erickson, is giving up control. Interactive writers are no longer telling their own stories but facilitating someone else's. They sacrifice the perfect pacing of scripted fiction for player agency: the ability to control one's own destiny. It is this feeling of player ownership that makes interactive fiction unique among genres of writing and that causes players to bond so intensely with their stories.

To find writers capable of telling labyrinthine, choice-filled stories in which the hero could be anyone, BioWare has created an intense vetting process. Candidates must submit writing samples and a fully playable quest in the Neverwinter Nights toolset to demonstrate their ability to write dialogue and nonlinear storytelling. If that submission is accepted, they repeat the process under a forty-eight-hour deadline, sometimes several times. Once they get aboard, writers undergo a three-month regimen during which they write training content; they are not allowed to work on the actual project until ready.

IT ALL BEGINS WITH THE WRITING

At BioWare, it was important that the writing team come together as quickly as possible. Because The Old Republic was to be voice acted and translated into multiple languages, the team would need a huge lead time to develop what would be enormous scripts. Plus, with BioWare's story-driven tradition, the locations, the characters—in short, *everything* waits for the writers. Building the team from scratch, Erickson reviewed thousands of candidates as he searched for writers who could deliver BioWare's nonlinear dialogues and plots.

An early Republic pilot sketch by Arnie Jorgensen.

STYLIZING REALISM

As the writers came aboard, one by one, Arnie Jorgensen and Jeff Dobson conducted early experiments with art styles. The first and most fundamental decision was that the game would be stylized. First, the game would be designed to be accessible for *Star Wars* fans with little or no gaming experience as well as for hard-core MMO veterans. Going for the most detailed, high-resolution graphics would mean the game would run at optimum capacity only on high-end computers and wouldn't look the same on all machines. Second, not only would it be years until the game launched, but years' and years' worth of subsequent expansion content would follow. An elegant, stylized look would look the same on computers of varying speeds and hold up aesthetically over time.

Most of all, the game needed to look distinct. "We wanted to react against the hyper-real that defines next gen," remembers Dobson. "I saw a lot of games that were very competently doing great shaders, super high-res stuff, but they all looked alike." Dobson and his team made a concerted effort to ensure that the *Star Wars* MMO, whether seen online, in print, or on television, could be nothing *but* the *Star Wars* MMO. Experiments with various types of stylization kept the team busy for several months. Jorgensen and Dobson explored a heavily pushed look, with muscle-bound comic book characters; visualized "dark age" Jedi with giant, unkempt beards and bare feet; and toyed with harking back to the original film by outfitting characters with retro clothing and hairstyles from the '70s. They finally decided on a unique, slightly exaggerated but extremely painterly look. "The decision to go for that hand-painted feel was a big part of us being able to define our own style," Dobson says.

Mood piece for the Jedi Knight by Arnie Jorgensen.

Once the final art style was determined, this Sith Warrior exploration was created as a touchstone.

CODE COMPLEXITIES

If programming games is difficult, programming BioWare-style RPGs is especially complex. Making the same style of game scale for thousands of players, keeping the frame rate high and, most important, guaranteeing that the game stays online and stable would be especially difficult for a company used to making single-player RPGs. Specialists were needed. Specialists like principal lead game-play programmer Damon Osgood and his counterpart, principal lead server programmer Doug Mellencamp. "A specialist that you can't find just anywhere in the games industry is a server programmer," says Bill Dalton. "When people think of games, they think of graphics. They think of what you see on the screen and how you interact with it. But for an MMO, you have a server. It's a vibrant, existing world that will persist and you need to be able to log into it and come back to it and find it in the same state as when you left—or at least your portion of it. The art to it is when you have things scale. When your code complexity gets up into the millions of lines and your systems have to be up for twenty-four hours a day, scale issues start to creep in, and you find yourself in a lot of trouble. That's where the engineering expertise comes in."

By 2006 both teams were in place and had a coherent vision for writing, art, and programming. It was time to start making a game.

2

BRINGING THE RPG TO AN MMORPG

"MMOs are very grindy, and you want to hide that grind. That's an advantage we have with story. You're constantly thinking 'What happens next in the story?' Not, 'I need to kill more zombies.'"

Brad Prince

BioWare
World Design Director

PAGE 28–29:
Early screenshot during art experimentation phase.

Various early sketches for Mandalore by Paul Adam.

A more developed sketch of Mandalore by Paul Adam.

Final concept art for Mandalore by Paul Adam.

The basic vision of TOR had been clear since the earliest meetings between LucasArts and BioWare: To make a full-fledged MMORPG in a way that had never before been attempted. MMOs grew out of classic role-playing game conventions. But the number of resources involved in creating these enormous games—games designed to hold millions of players' interest for months on end—meant that most titles were often stripped-down versions of single-player RPGs, stretched thin to hold hundreds of hours of very basic content. Intricate stories and branching quest lines—a hallmark of RPGs—had never made it into the MMO space. Nor had fast-paced combat. "Maybe ironically, we did not set out to create a very different combat system. But we did have a set of prerequisites: The combat had to look like it belonged in the *Star Wars* universe, with lots of lightsaber clashes and laser bolts; the combat had to make the player feel heroic; and the combat in itself had to be fun, since players would be spending a good chunk of their time fighting," says Emmanuel Lusinchi. "It was only later on, after most of the combat systems were completed, that it became clear just how far from many other MMOs our combat and story experiences ended up."

Khem Val, a dashade companion character by Arnie Jorgensen.

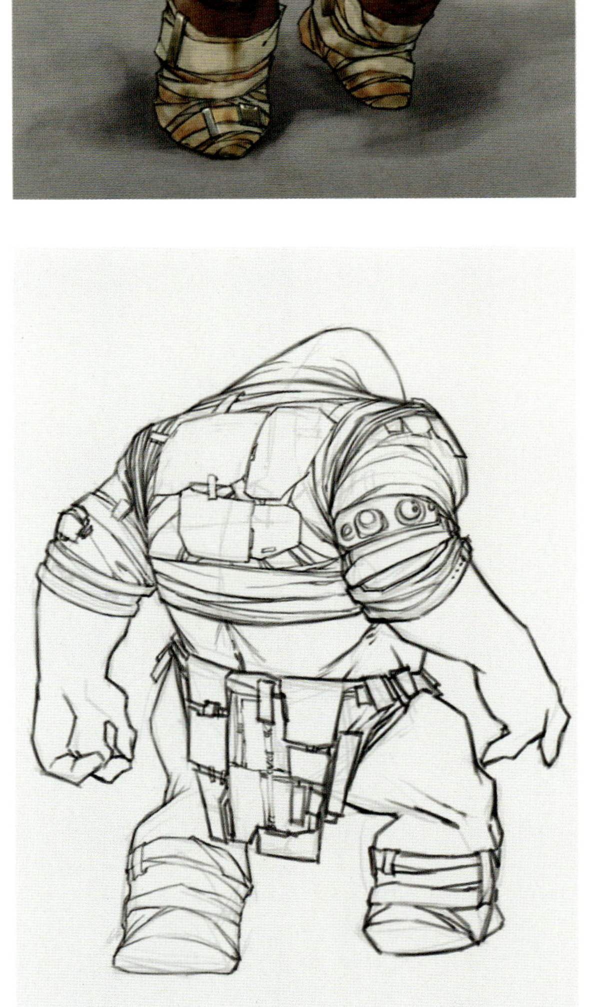

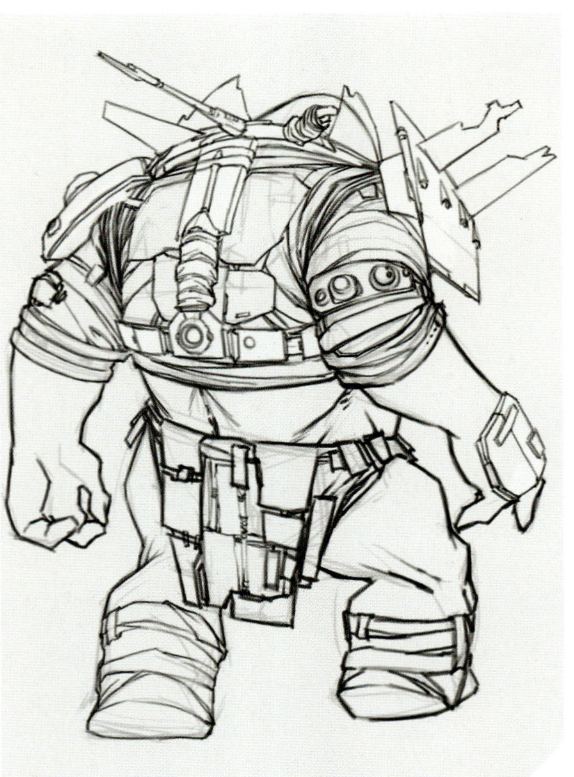

One of many flesh raider variations done by Arnie Jorgensen.

Broonmark, a talz companion character by Arnie Jorgensen.

ROLE-PLAYING GAME PILLARS

"A successful role-playing game has four pillars and does them all very well," says James Ohlen. "Exploration, combat, advancement/customization—and story." As the first RPG pillar, exploration was easy to picture, though it would take years of work to pull off. *Star Wars* comprises a massive galaxy, with huge distinct worlds, countless crazy aliens, and thousands of years of history. Players would be free to travel among at least a dozen worlds that strike a balance between KotOR legacy locales, such as Korriban; famous worlds from the movies, like Hoth; Expanded Universe favorites, like Alderaan; and new additions to the galaxy, like Voss.

From the onset LucasArts and BioWare agreed: Combat needed to be fast-paced and immersive, like a high-action console experience, and more compelling than any other MMO to date. "Given that action-packed combat is a critical element in the *Star Wars* films, the team knew from the outset that it would be important to offer players of The Old Republic compelling *Star Wars* combat," says Leo Olebe, BioWare's director of marketing. "MMO consumers have been wanting some game to break away from stilted hack-and-slash battles and instead provide an action-oriented experience." In the end, combat would become one of the most complex parts of the game, eventually involving a unique cover system.

When it came to customization, Ohlen aimed high from the very beginning. Each class would need its own set of companion characters, its own ship (though the Jedi and Sith classes would eventually share this resource), and a huge number of custom assets and weaponry, including more class-specific gear than many games had total outfits. The artists would need to create an enormous variety of clothing (wearables) for all of the player classes while ensuring that all wearables remained true to each class at all levels.

The first concept ever done for the game by Arnie Jorgensen. Used only for internal promotion.

THE FOURTH PILLAR

By their very nature, traditional MMORPGs contain plenty of the first three pillars. But none had truly incorporated the fourth: story. Doing so would be completely new in the MMO space. Using classic storytelling techniques meant telling a story over time, introducing interesting characters, giving players a chance to bond with them, then giving players choices about how to interact with those characters, which would then change the path of the story forever. "Story is what brings an RPG together in an emotionally powerful way," says Dallas Dickinson, BioWare Austin's director of production.

Dickinson adds that the decision to add the story pillar to the game was a goal from the beginning, designed to appeal to players of all backgrounds and levels of video-game experience. "MMOs can be a bit intimidating to some players, but a great, story-driven RPG—especially one based in the *Star Wars* universe—is way more compelling and accessible. Players who have dreamed of a great *Star Wars* story where they are the hero are going to have the chance to be that hero in The Old Republic."

In an MMO, players engage in activities integral to progression, such as combat, group play, questing, and crafting. All these activities can trap players in a grind of killing as many enemies and snatching as much loot as possible to build their characters. Adding story would give players context and slowly introduce them to these activities and the other three game pillars in an appealing and easy-to-understand way. "The game's story is driving all of that stuff," says Neri. "You're participating in your story, you're making decisions that are critical to you as a player, and that sends you out into the world to do other activities that you would do in an MMO."

"No other MMO has really explored having a meaningful story to support why you're going on quests, why you're fighting a different faction, why you even want to level up," says LucasArts senior product marketing manager Rob Cowles. "With The Old Republic, story actually gets you more immersed and gives you a purpose for doing every activity in the game."

Blizz, a Jawa companion character by Diego Almazan.

An original piece done for early marketing by Arnie Jorgensen shows a Jedi with HK-47.

The same piece changed to show a Sith instead of a Jedi.

CHOOSING BETWEEN LIGHT AND DARK

One of the most popular features of Knights of the Old Republic was its system of assigning light- and dark-side points for choices a player made in-game. It was an immediate commentary on the player's actions, fit perfectly with *Star Wars*, and was a no-brainer to continue in The Old Republic. Taking the system directly from KotOR, however, turned out to be untenable. One of the critiques leveled at the original system was that a light-side player had to be ultrapolite, as even harmless jokes and off-color remarks would sometimes result in dark-side points. It was a minor nuisance, but it caused some players to reload saved games to make sure they were seen as the paragons or villains they intended. Being an MMO, The Old Republic would have no save and replay features. Once made, choices were permanent. "You have to remember that every decision you make can have an impact on your game-play experience. The impact may be immediate (the captain is dead!) and others may not be felt until you are much deeper into your personal storyline," says Blaine Christine, senior producer at BioWare. As in all BioWare games, these decisions are rarely easy and require the player to make moral decisions that ultimately inform them about the character they've chosen to play in the world of The Old Republic—and perhaps even about themselves.

Taken into the class-specific stories of The Old Republic, the system was even more problematic, as the situations demanded more subtle behavior. While playing a light-side Sith was certainly possible, it made no sense that those players would announce their light-side ways to their Masters. Similarly, corrupted Jedi needed to be able to maintain a plausible cover. TOR's system therefore became purely action based, reacting to what players do instead of what they say. A young Sith can look his Master straight in the eyes and promise that he will go commit a massacre—but it's when he instead secretly prevents that same massacre that points are awarded.

The effects of dark side corruption by Arnie Jorgensen.

An early sketch by Paul Adam from the development of Darth Jadus.

FROM SILENTS TO TALKIES

The final major way in which the game would progress the genre was in the area of voice-acting. In a typical MMO a player clicks on a character and receives a block of text to read and a quest that might have little or nothing to do with their character. TOR would feature cinematic, voice-acted conversations that would create context and emotion, forcing players to define themselves through their actions. Choosing to be greedy or valiant, dismissive or empathetic, vicious or merciful would come back to haunt or reward players further on in their stories. With full-voiced dialogue, characters in TOR would be living, breathing people with motivations that were not always clear, who could betray or befriend the players and form real relationships with them.

GETTING THE SYSTEMS IN PLACE

The development team started to build the core systems of the game in early 2007. During stage one the designers came up with a prototype of melee, blaster, and force combat, and created basic characters and creature movement. While simultaneously focusing on the development tools, the team then added extra systems, such as the conversation system, maps, journals, and social systems, and started fleshing out advanced combat ideas. Once the core systems were tested, development tools were designed for the programming team and production tools were created for the designers and artists who would need to build thousands of hours' worth of content.

Next, the team started working on complicated technical problems, such as how *instances* were going to work. (Instances are spaces in the game where the players can't see each other and can have their own experiences or play with just their own group.) Early technology development was a balancing act and was often two steps forward and one step back. The team couldn't follow the normal non-MMO development process of first building a prototype and then creating content. Prototypes had to be built and rebuilt while tools and revisions to those same prototypes were being created simultaneously.

3

THE FUTURE OF AN ANCIENT TIME

"I remember having a distinct conversation with Ray and saying, 'Some of the fan base is asking "What about Knights of the Old Republic III or IV?"' And his response was, 'Because of the vastness of this world and the multiple choices people have and the different character classes and the different worlds, this is in essence Knights of the Old Republic III, IV, V, VI, VII, and VIII.'"

Mary Bihr

LucasArts
Vice President of Global Publishing

PAGE 40–41:
An early exploration of visuals for the "Deceived" trailer by Blur Studios.

By 2007, the game's artists were busy churning out conceptual art while the writers, guided by James Ohlen, began constructing the overarching story against which the game would take place. It was important that the established events and mythos of the Old Republic era be maintained while giving the writing team room to work. As it became clear that TOR would have more story content than all other Old Republic sources put together, BioWare and LucasFilm decided to place the game three hundred years after the events of Knights of the Old Republic II. While a blink of the eye in terms of technological or historical changes in *Star Wars*, the three-century gap nevertheless was enough time to put whole new plots in motion and new actors on what would need to be a very large stage. "The great thing about *Star Wars* is that there's all kinds of lore that does go back that far," says Muzyka. "What we endeavored to do was find the lore that fit and put it together in a way that made sense for an MMO."

TOR would contribute a huge chunk of new continuity to the *Star Wars* mythos, and given *Star Wars*' science-fiction milieu, creating detailed setting would provide much-needed verisimilitude and depth. "This may sound silly, but one of the key things we find is that you need to have a really good understanding of the political environment, the cultures," says Zeschuk. "It's not just about the pretty pictures and the environments, it's about the people and what they're doing and why they're doing it. And that sets the stage for the whole conflict that the game covers."

Concept art of Eleena created by Petrol during preproduction for the "Deceived" trailer.

HISTORY OF THE OLD REPUBLIC ERA

Sith mood piece by Arnie Jorgensen.

Those familiar with the interwoven tapestry of tales that exist outside the *Star Wars* movies (known as the Expanded Universe, or EU) may not appreciate how much of a creative leap it was to stray thousands of years before the time of Skywalkers and Solos. Dark Horse Comics took the leap in 1994 with *Star Wars: Tales of the Jedi*, a series that immersed readers in an archaic, almost mythical era of the *Star Wars* galaxy. Subsequent volumes of *Tales of the Jedi* cumulatively chronicled the rise and fall of the ancient Sith Empire, the Great Hyperspace War, and the Republic's earliest struggles against the Mandalorians.

Video games would be the next medium to flesh out the era. Knights of the Old Republic hurled players into a tumultuous Republic contending against a military conquest led by Darth Malak and Darth Revan. The following year LucasArts and California-based Obsidian Entertainment released Knights of the Old Republic II: The Sith Lords, in which the Jedi Order has been eradicated by the Sith, who dispatch dark-side assassins throughout the galaxy to locate an exiled member of the Jedi Council. Dark Horse's *Knights of the Old Republic* comic series (2006–2010) bridged the gap between KotOR and the *Tales of the Jedi* comics preceding it, depicting the horrors of the Mandalorian Wars while setting the stage for Malak and Revan's war against the Republic.

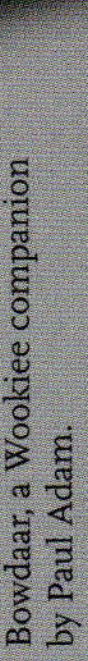

Bowdaar, a Wookiee companion
by Paul Adam.

As The Old Republic opens, the galaxy is in the latter stages of recovery from a horrible war. The original Sith Empire, chased out of known space a thousand years before, returned and exacted devastating revenge upon the Republic. Led by a mysterious Emperor rumored to be hundreds of years old, the Sith ravaged planets, destroyed fleets, and sacked the Republic capital of Coruscant. Then, when it appeared there was no hope, the Emperor had a seeming change of heart and offered a peace treaty in exchange for uncontested ownership of several small and apparently worthless systems. In no position to argue, the Republic signed the Treaty of Coruscant, and for a dozen years a strained peace held. Now, however, border disputes are the norm, Republic ships are disappearing, and the galaxy seems to be sliding toward war with heroes lining up on both sides.

A very early Korriban sketch by Arnie Jorgensen.
The bridge was eventually removed for gameplay reasons.

The very first Hoth concept by Arnie Jorgensen
already featured the ship graveyard.

IT'S CALLED STAR **WARS** FOR A REASON

While planning out the events of The Old Republic, the writers followed a few overriding principles. First, the game needed to feature an epic, galaxywide conflict that was easily understood: the war between the Republic and the Sith Empire. "*Star Wars* is all about big, ridiculous threats. You have to have your Death Stars, your planets exploding, your giant fleet battles, and huge ground battles," says Ohlen. "For the duration of our MMO, we wanted to make sure that there was always a war." Because the game would enable players to take either side of the conflict, neither side could be a mystery or unveiled as the game progressed.

Second, the story events needed to fit existing *Star Wars* lore and bridge the gap between the Knights of the Old Republic era and what would come after. Third, the conflict had to be at a low point when the game started. It takes players time to get grounded in a new world, to understand their role and what's going on around them. Starting players right before the Old Republic equivalent of the World War II invasion of Iwo Jima would cause confusion and chaos. What the game needed was an equivalent to the days just following Pearl Harbor, when two sides were looking at the inevitable but had not yet seriously clashed.

This early sketch by Arnie Jorgensen was used to show a rougher, more ad-hoc approach to personal ships.

Early Imperial soldier helmet detail by Arnie Jorgensen.

SITH TROOPER

Early Imperial soldier design by Arnie Jorgensen. Both the look and use of the word “Trooper” on the Imperial side were later abandoned.

Imperial soldier design by Blur Studios.

Fullsheet concept by Paul Adam illustrates differences between Imperial soldiers, heavy Imperial soldiers and Imperial Guard.

Early sketches for high-level Sith Warrior gear by Paul Adam.

Various attachment options for one set of Sith Warrior gear by Paul Adam.

THE TRUE NATURE OF THE DARK SIDE

Most role-playing games assume that players are heroes working for law, order, and the status quo. The Old Republic would be unique by allowing players to promote darker ambitions as members of the Sith Empire. To make the Sith easily understood, the writers went back to the original Sith Empire, which was chased out of known space following the Great Hyperspace War. Since that time there had been many people calling themselves Sith, but many were fallen Jedi, members of odd one-off cults, or other complicated plot devices. Teaching new players that they were Sith but they actually used to be something else sounded difficult to communicate.

And it was far less dramatic than being purely Sith. The original Sith Empire had things to separate it recognizably from the Galactic Republic. Except for the red Sith-blood races, it was entirely human, while the Republic was a mishmash of different species. The Sith had their own culture and forms of government. Most of all, establishing a working empire for the Sith made them people instead of monsters, with a distinct point of view that Imperial players could understand. "It's very difficult to portray a realistic society that thinks of itself as evil," says Ohlen. "We tried to make the Sith Empire as realistic as possible and a place where people would actually serve. But at the same time, we had to make sure that the rulers of the Sith are very evil people who do horrible things. You are sent on missions that further the Sith cause, and the missions are inherently not nice."

Because *Star Wars* fans know that the Sith Empire does not exist centuries later, the writers also had to lay the groundwork for how the two governments would eventually merge, taking pieces of later Galactic Republic culture and saying "these bits actually came from the Sith Empire."

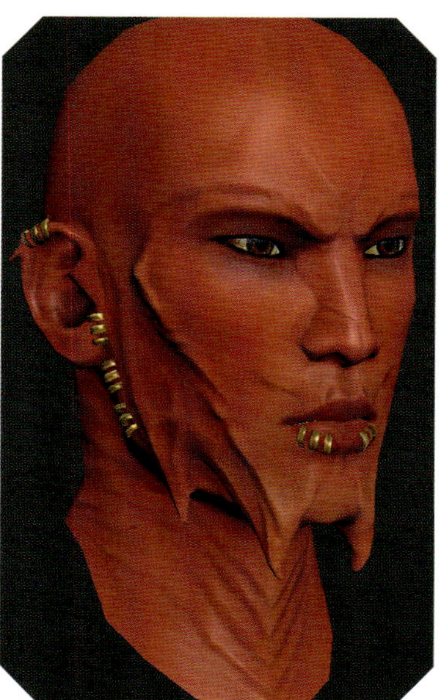

In game model of Pureblood Sith female jewelry by Sperasoft.

MORBID SECRETS

Concept of the Emperor's throne room by Paul Adam.

There needed to be questions for players to solve as well, so the writers focused on the Emperor as the mysterious mastermind behind the invasion that neither side understood. "No one has seen the Emperor since the Treaty of Coruscant," says Hall Hood, the senior writer in charge of the Jedi Knight story. "He rules his Empire through proxies, but even they don't know their master's true plan."

The Sacking of Coruscant explained why the Jedi would sign a treaty with their hated enemy—creating the lull in the action that the writers wanted in order to bring players up to speed—while it also built resentment against the Empire and provided the first great mystery: Why had the Emperor, when he seemed to have victory in his hands, stopped the invasion in return for several worlds of no apparent strategic value?

4

CERTAIN POINTS OF VIEW: CHARACTER CLASSES

"I've been trying to make massively multiplayer games for twenty years. There is a very real problem in actually having a heroic arc [in which you] can grow from being a fledgling to fighting truly epic battles against epic adversaries. In *Star Wars* you start as a Padawan, you grow, you fight bad guys, they feel tough, you feel heroic. By the end of the experience you're fighting AT-ATs and rancors and these other big experiences that come from the *Star Wars* universe. *Star Wars* is about the only thing that even comes close to fantasy in terms of providing a great heroic arc."

Damion Schubert

BioWare
Principal Lead Systems Designer

Page 54–55:
Class concept pieces for the Jedi Knight, Bounty Hunter, Jedi Consular, and Sith Inquisitor by Sperasoft.

Once the galaxywide conflicts and timelines were in place, the writers and designers turned their attention toward creating the player-character classes. "In Dragon Age we had these one- to two-hour class-specific openings called origin stories, which were extremely popular," Daniel Erickson recalls. "We thought, *Wouldn't it be amazing to do these throughout an entire game?*" Consequently, every class in the game would get its own custom story that stretched from level 1 to the end of the game.

The implications of this decision quickly became apparent as it meant each class would need its own set of companion characters, its own ship, and a huge amount of class-specific gear. Limited by the work that could be done in the years allotted, the team limited the number of classes to eight. Every class, however, would be split down the middle to create two distinct "advanced classes" for each, from which players could choose. This option gave the game sixteen different classes from a balance and game-play perspective while requiring the team to write and record only eight separate storylines.

When developing the class concepts for the game, the team looked at them from two angles: What fans of the *Star Wars* movies would want to play and what fans of the MMORPG genre would expect. "A lot of times when a team starts to design an MMO they say, 'OK, we're going to make our classes. We gotta have a tank, we gotta have a healer, we gotta have a DPS [Damage Per Second, a class that focuses on causing quick damage],'" says Jake Neri. The way the designers approached The Old Republic was much different. "We wanted to integrate both sets of expectations tightly into the class/advanced-class design," says Georg Zoeller, the game's lead combat designer. "The result is a lot of versatility per class with many options outside the traditional [healer/tank/dps] trinity of MMORPG classes."

"We just said, 'OK, screw the RPG tropes for now,'" says James Ohlen. "Let's just think of what people actually want to play in the *Star Wars* universe." Using the *Star Wars* movies as inspiration, the class archetypes were cut-and-dried: Palpatine would be the basis for the Sith Sorcerer (which later became the Sith Inquisitor); and Darth Vader, the Sith Warrior. Luke and Anakin Skywalker inspired the Jedi Knight; Yoda became the Jedi Wizard (eventually renamed Jedi Consular). The teams had Han Solo in mind when conceiving of the Smuggler, Boba Fett for the Bounty Hunter, and the clone troopers for the Trooper class. "Regardless of what role the player needs to play in a group and in the game, it needs to feel like the movies," says Neri. "The movies are what people know. If we're not delivering that, then we're totally failing."

Erickson recalls the writers struggling to come up with a player class that would represent the Empire and yet offer something new to *Star Wars*. "We didn't want to do another soldier and said, 'What's going to make it more about the Empire?'" says Erickson. "The Republic Trooper can do things that the Jedi won't do for moral reasons. There is nothing that the Sith won't do, so who do you need?" And so was born the Imperial Spy class, later renamed the Agent. That the team would write a James Bond-esque spy story was a foregone conclusion until senior writer Alex Freed came up with an alternate idea. "Alex came back to us and laid it out: 'No, no, no, no. We're going to do a *24* piece. A *Bourne Identity* piece. This is the Empire. It's serious. It's dangerous. You've got a lot of very hard decisions to make.'" Though there would be elements of a classic spy thriller, the Agent story would be gritty and intense. "Alex was right about the direction that it needed to go, and he turned in one of the most amazing scripts I've ever seen," says Erickson.

Though the Agent class didn't correspond perfectly to a famous *Star Wars* hero or villain, certain elements of intrigue from the films—Lando and Leia's infiltration of Jabba's palace, Darth Maul's tracking technology, and Bothan spies—provided precedents. "You'll see lots of familiar things," says Freed. "Alien informants, political assassinations, sabotage missions, 'techno-logical terrors,' the Agent focuses on all the espionage scenarios and cool gadgets you see in the movies."

Other classes that were on the board but that didn't make the final cut were the Jedi Warden and the Sith Assassin, both of which became advanced classes (albeit under different names); and the Mystic, a gray-aligned Force user who was neither Jedi nor Sith. Eventually, the Mystic class morphed into the Voss species.

Early clothing design by Arnie Jorgensen.

One piece of a large mural painting showing the character classes by Marek Okon.

A second piece of the same large mural painting by Marek Okon.

FLESHING OUT THE CLASS STORIES

The writers then went back and built out the galaxy to support the classes. If they were going to write a Trooper story, they first needed to figure out everything about the military on both sides: jargon, ranks, military life. It became necessary to keep living documents that tracked the Imperial and the Republic military structure, history, and major battles so that other writers could present any Trooper content with a consistent level of detail. The creation of the Imperial Agent class spawned the creation of not just Imperial Intelligence but the Republic's version of America's Central Intelligence Agency, working on the other side of the spy game.

Full of action and adventure with a healthy dose of humor and romance, the Smuggler class story probably couldn't exist as its own RPG. "The Smuggler was the one that surprised everybody, because we got to do a light action comedy," says Erickson. "We started to ask, 'What makes a Han Solo adventure?' Awesome lines, always being in over your head, everything totally goes wrong, and as many ridiculous attempts at romance as possible.

"The Bounty Hunter [story] is almost a western: You are your own man trying to make ends meet," says Erickson, who penned the first worlds of the Bounty Hunter class story. "There's a whole bunch of bad hombres out there who are going to try to take you down. You can only count on the people around you." The Bounty Hunter is also the only class on the Imperial side that isn't strictly Imperial. As a hired gun, the player has an outsider's view and can make autonomous decisions about employers.

For the Smuggler and Bounty Hunter classes, the underworld needed to be fleshed out and vibrant. The EU fan base—and many of the writers—had a love affair with the Mandalorians, so it was important to know what they were doing during all this as well. At the end of preproduction, there were more than a thousand pages of the design dedicated to story and background. Each faction had its own document with intricate details on its history and relationship with every other faction. Muzyka stresses the importance of so much preproduction world building: "There are aspects that are shown to players explicitly and there are aspects that are beneath the surface. It adds a certain weight, a certain credibility, a certain reality to what is above the surface. It has that sense of being real. It feels connected, as if it's part of something larger."

Finally, to ensure that all classes would perform equally well in the game space, the designers determined the abilities of each class and how those abilities would actually play. Feeling that it would reduce player options and limit group play, they avoided shoehorning the classes into MMO types. Instead, they gave each class a specific play style: ranged—or distance—weapons, melee weapons, lightsabers, Force powers, stealth—with enough flexibility to fulfill different roles in a group. It enabled the designers to break the mold of traditional MMO archetypes whenever they could. "We didn't set out to make the Sith Warrior into a traditional melee tank. Instead, we asked ourselves 'What would Vader do?' and built from there," says senior game designer William Wallace. "The way each class plays developed organically from the actual character archetype, not from trying to fit traditional game mechanics to the classes."

Smuggler and Wookiee companion character for magazine cover from outside marketing firm.

Female bounty hunter concept by Sperasoft.

Darth Malgus preproduction design by Petrol.

Darth Malgus 3D render by Blur Studios.

BRINGING BALANCE TO THE FORCE

It was crucial that the game balance Force-using and non–Force-using classes so that players wouldn't gravitate toward the former overwhelmingly. As non–Force-users, the Trooper and Bounty Hunter would have access to armor and powerful weaponry, such as missile launchers and thermal detonators, and could attack from a distance. "Jango Fett managed to fight Obi-Wan to a standstill in *Attack of the Clones*, so I think people are willing to accept that a Bounty Hunter could fight against a Jedi," says Ohlen.

Ensuring that the Imperial Agent and the Smuggler were just as competitive was another challenge. "The Smuggler is based on Han Solo, and the only time he ever goes up against a Sith is against Darth Vader in *Empire*, and he pretty well gets his ass handed to him," says Ohlen. Because they didn't want players to be constantly trounced by Sith and Jedi, the game designers needed to ensure that Smugglers and Agents could beat the ostensibly more powerful classes in ways that would be logical to players. So cover, stealth, and gear were made part of the Smugglers' and Agents' repertoires, resulting in complex classes that use tactical approaches to devastating effect.

Female Bounty Hunter 3D render by Blur Studios.

As an already tenuous peace starts to unravel and full-on war strains the seams, Jedi Knights, caught in the eye of the storm, grow in ability from novices to full-fledged Masters of the Force. The class designs would reflect this progression. Inspired by Obi-Wan Kenobi's armored look in Genndy Tartakovsky's *Clone Wars* microseries, the line for the Knight class consists of muted tones and cloth robes draped over heavy, plated armor, some suits of which are etched with mystical circuitry and symbols. "The Knight goes from a young Padawan look, with brown robes, to one of the more interesting looks in our game, which is the whole ancient techy armor look, which incorporates old sigils and runes," says Arnie Jorgensen, who created the look and feel of the eight classes.

"Lots of *Star Wars* clothing without a techy belt or a bandolier with chunks of metal on it could very often pass as medieval," says Jeff Dobson. "Jedi without lightsabers look like monks. Luke in Episode IV is in a smock, and if he doesn't have a techy belt or isn't driving a landspeeder, he could easily be in a King Arthur movie and not look out of place. It's an interesting dichotomy with *Star Wars*. It's simple, old fashioned-type stuff mixed with the future. Big, simple shapes and sometimes fills of very high-frequency patterns."

Jedi Knight wearables by Arnie Jorgensen.

Lightsaber hilt designs by Sperasoft.

Jedi lightsaber hilt design by Sperasoft.

Jedi Knight design by Petrol.

Whereas the Jedi Knight leans toward combat-readiness, "the Consular looks more like a wizard," says Jorgensen. "They have the ancient, ruined look of the old Tython." Jedi Consular concepts were some of the earliest that the artists explored. *Star Wars* is a marriage of science fiction and fantasy; to articulate this, Jorgensen originally melded sci-fi with the classic "wild man in the desert" look. Consulars sport animal hide and hand wraps, archaic tech, and tattered robes edged with strange, glowing runes. "Oddly enough, the thing that still comes to mind is a goofy little pencil sketch I did of a wizened old Jedi early on in the game," says Jorgensen. "That look was enough to inspire the entire line of the Jedi."

The final design for the Jedi Consular class would move away from Jorgensen's original ideas, however. The concepts would become more elaborate: flowing, ornate robes, bold shapes with intricate patterns, and exotic ceremonial headdresses. The original designs were eventually incorporated into the game as high-level Player versus Player (PvP) gear.

Jedi Consular PvP gear by Arnie Jorgensen.

Maps for re-coloring Jedi Consular gear by Diego Almazan.

Republic Trooper concept design by Paul Adam.

HEAVY METAL: TROOPER

In designing the Trooper class, the artists drew on prequel-era clone troopers; classic-era stormtroopers, snowtroopers, and scout troopers; and even Rebel soldiers and X-wing starfighter pilots. "We went back to the original series and looked at the trooper armors and the plasteel look; we wanted to bring that into the Republic Trooper," says Jorgensen. "The most interesting thing to me about *Star Wars* is the '70s boxy tech. We tried to keep it in the game, because that's what helps distinguish it from other sci-fi."

Outfitted in camouflage, the jungletrooper look was based on that of the clone troopers who fought alongside the Wookiees in *Revenge of the Sith*, while the snow variations were inspired by unused Ralph McQuarrie concepts from *The Empire Strikes Back* (albeit much more imposing). Trooper armor implies function, with lots of gear and retro design elements incorporated into the mix. Communications devices, helmets, antennas, visors, and shoulder pauldrons combine with big flashing buttons and blocky shapes. Alternate designs look hand-painted and grungy, implying heavy wear and tear, while others sport painted-on fangs and bristle with deadly looking spikes, lending a more aggressive look.

Small images by Paul Adam.

Large image by Sperasoft.

Smuggler wearable by Ben Huen.

Smuggler concept by Sperasoft.

Various smuggler blasters by Sperasoft.

FROM DROMUND KAAS WITH LOVE: AGENT

Agent concept by Sperasoft.

"The Imperial Agent was a rough one, because he isn't in the movies," says Jorgensen, whose challenge was to design a character class that was Imperial, but which could not *look* Imperial. A spy can't give himself away, after all. Many of the Agent designs were thus designed to convey intrigue and clandestine acts: lots of cloth and leather. Trench coats with high collars and shades give a sense of cloak-and-dagger coolness, while pitch-black body suits and feature-obscuring hoods imply espionage and assassination. "The Agent is acrobatic and ready for action. It's more of a psychological operations-type look. He's the sniper that's going to kill you."

As in the Trooper class, everything the Agent touches has function, only it's all smooth and economical with lots of modern tech, cool gadgets, and compartments for concealed items. Even their firearms are elegant. The class would also need a wide array of costumes for infiltration missions requiring deep cover. "He's got a lot of dark wearables and some wearables that can be a variety of things," says Jorgensen. "All the way from underworld up to Republic and Imperial as well."

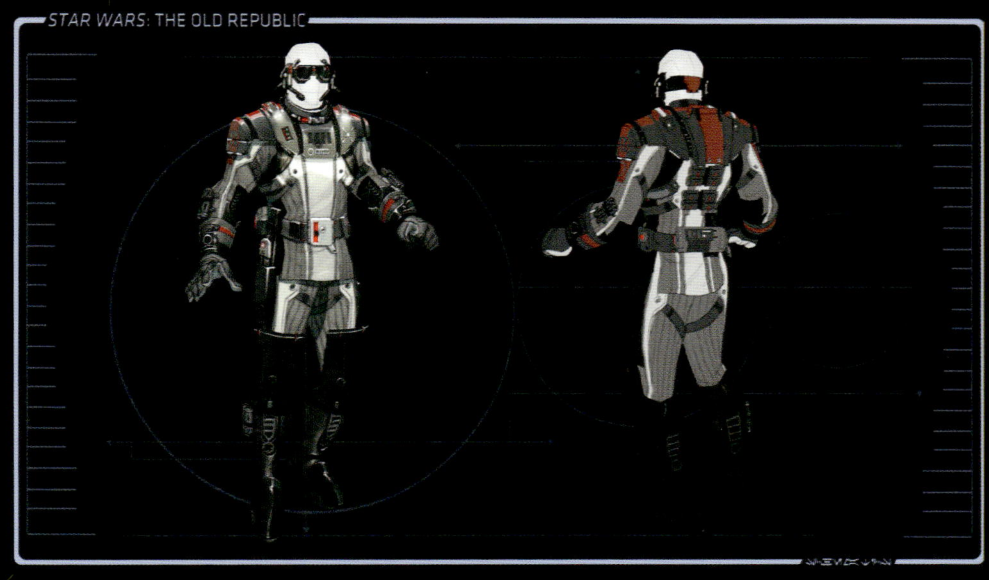

Agent wearables by Ben Huen.

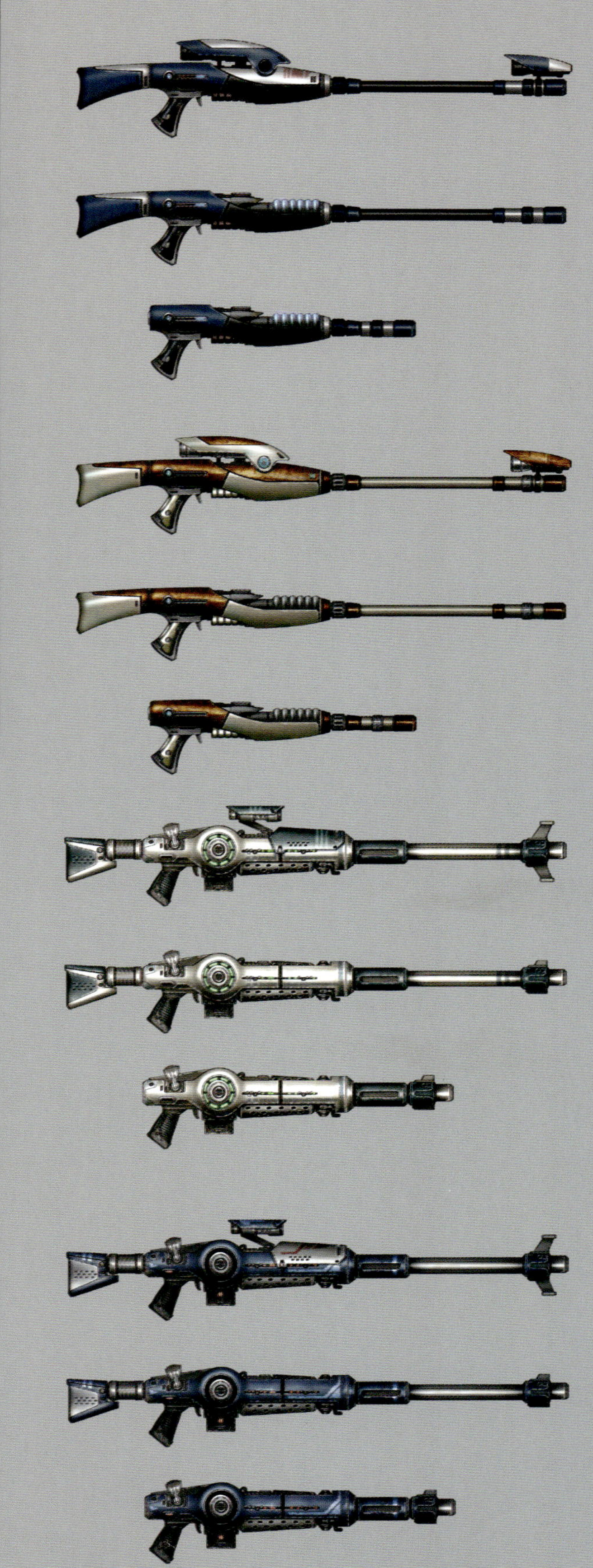

Various Agent blaster rifles by Sperasoft.

Bounty Hunter concept by Sperasoft.

THE MOST DANGEROUS GAME: BOUNTY HUNTER

Though Boba and Jango Fett served as the main inspiration for the Bounty Hunter, game artists explored a wide array of looks for the class. Random denizens spotted on Mos Eisley in *A New Hope* and bounty hunters from *The Empire Strikes Back* all provided visual grist, as did Mandalorian warriors. Because the Bounty Hunter needs to survive a dangerous profession in an even more dangerous galaxy, the line look needed to feel battle-worn and improvised. That meant weathered, lived-in armor with exposed cloth ribbing, mixed and matched pieces swapped from other suits, and lots of accoutrements: tubes and hoses, equipment that lights up, bandoliers, pouches, wrist-mounted flamethrowers, and thermal detonators.

In the same way Boba Fett wore braided Wookiee scalps, other designs include similar souvenirs from various bounties: feathers, teeth, and claws. "We're trying to hit all the main notes of the bounty hunters that we had seen in the movies," says Jorgensen. "The cool thing is that these guys are armored gadgeteers. They're not uniform."

Various Bounty Hunter face paint options by Arnie Jorgensen.

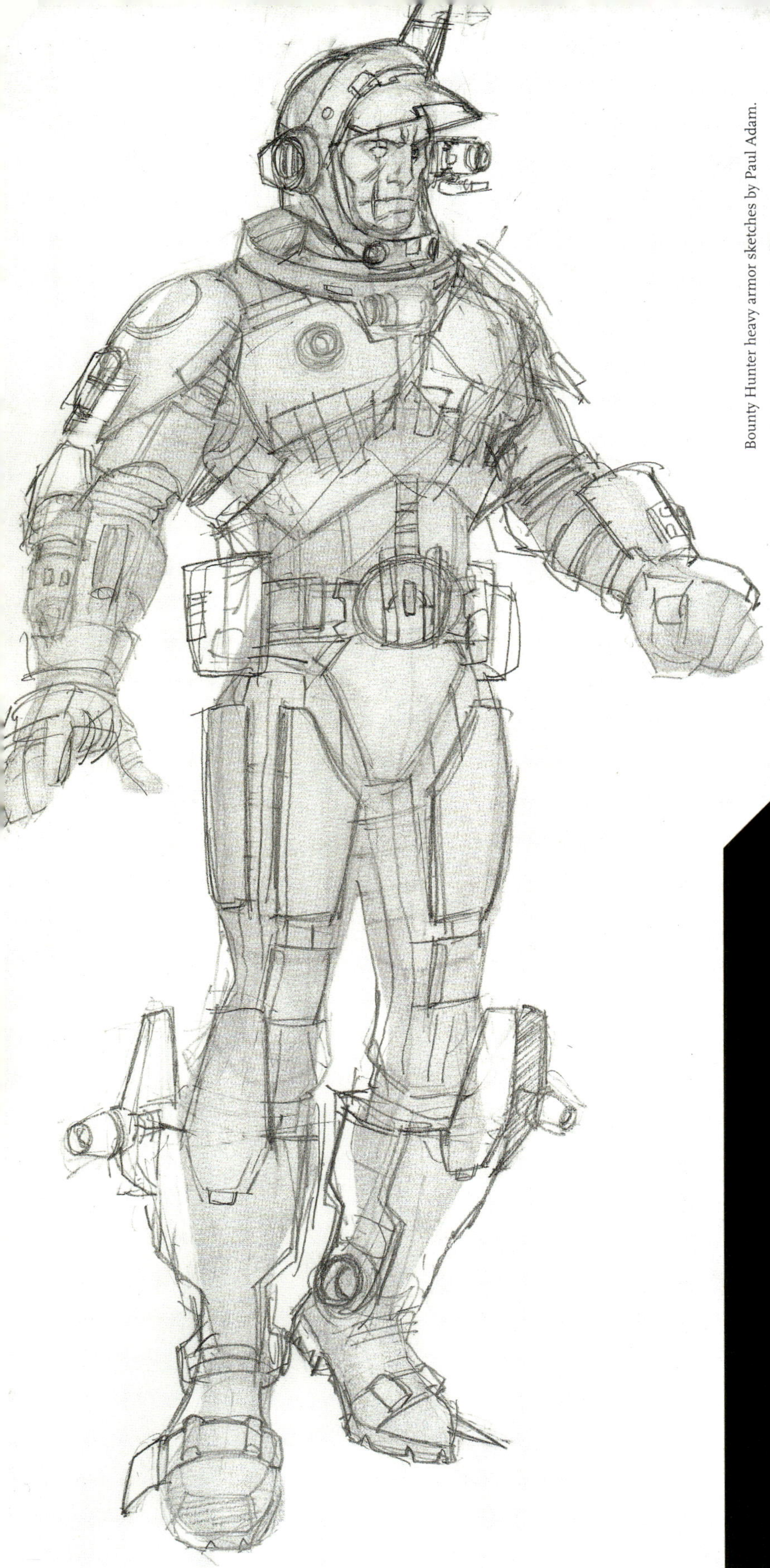

Bounty Hunter heavy armor sketches by Paul Adam.

Bounty Hunter wearables by Arnie Jorgensen and Paul Adam.

nor design by Arnie Jorgensen
Malgus's look.

DARKNESS: SITH WARRIOR

With armor integral to the class, many Sith Warrior designs are harrowing, machinelike permutations of Vader: lots of hoods, capes, chest plates, and control boxes. Whereas Troopers and Bounty Hunters appear protected in their armor, Sith Warriors look like walking torture devices, with caged face masks, piercings, protruding edges, breathing apparatuses, and clawed toes—all of which look painful to wear. Some of the more pushed designs border on mech, complete with servos and quasi-supernatural tech, implying bodies so twisted by the dark side that they require artificial means of support. "The Sith Warrior goes into some really interesting realms of evil cybernetics with spikes and crazy stuff," says Jorgensen. "They're very heavily armored, and that armor is built for fear. They have lots of dark metals and claws and runes on them."

The team explored other options by imagining Vader in a number of hypothetical scenarios—sans helmet, in full armor, with cybernetic equipment replacing the quilted leather of his suit—and also drew on the Emperor's royal guard and some discarded Sith character concepts from the films. "There's a lot of great stuff in the prequel *Art of* books, rejected concepts, and that's a gold mine," says senior concept artist Paul Adam. "There's really great Asajj Ventress artwork. There are also lots of revisions of Vader and other villains throughout the series that never actually made it into the films, and we borrowed from there as much as we could."

Sith lightsabers by Clint Young and Arnie Jorgensen.

Sith Warrior design by Paul Adam.

ANGEL OF DEATH: SITH INQUISITOR

Darth Jadus concept by Paul Adam.

To lend the Sith Inquisitor class a horrific feel, the artists took cues from Wayne Barlowe's artwork in *Barlowe's Inferno* and from the sadomasochistic Cenobites in the *Hellraiser* films, marrying them with dark mages from classic fantasy. Smoke and dust seep from tattered robes that verge on disintegration and which are adorned with organic elements associated with decay and death: leather husks, mummified carapaces, and fossilized remains of strange creatures. Arnie Jorgensen describes the class as "nightmarish. They're like a royal look in the Sith hierarchy. They can also level-up to twisted Force-powered armors that look more evil and skull-like."

To further distinguish the Inquisitor from the Warrior class, the artists focused on color palettes of black and violet, with exotic accoutrements: protruding spikes, deranged masks topped with freakish plumage, and horned helmets. Because the Inquisitors serve both the Sith and Imperial causes, the designers also imagined they'd be willing to incorporate technological elements into their design. "If we take a robed evil sorcerer and allow him to tech himself out," says Adam, who oversaw player-class wearables, "and then take that further and corrupt it with the dark side of the Force and twist it and make it look evil—that's where we get our Inquisitor."

Inquisitor outfit concepts done by various artists.

Inquisitor concept by Sperasoft.

"When you get in the game and you walk outside for the first time, you realize that you're in this gigantic *Star Wars* universe on this huge planet. Not only are you going to be running around this planet for hours, [but] knowing that there's a dozen other planets to explore is overwhelming. Just the scope of the exploration is going to be hundreds of hours of *Star Wars* geekdom right there."

Orion Kellogg

LucasArts producer,
online

Page 84–85:
Operation: Eternity vault by Ryan Dening.

With hundreds of worlds to choose from in the *Star Wars* galaxy, the writers and designers followed a number of different criteria when deciding which ones to include. First, they wanted a rich variety of planets that encompassed recognizable worlds from the *Star Wars* films, little-seen worlds from Expanded Universe literature, key locales from Knights of the Old Republic, and uncharted territory. Second, each of the worlds needed to be visually unique. It played into *Star Wars*' tradition of planets dominated by one ecosystem and color palette. Since players would be spending hours on each world, it would also provide much-needed variety. Finally, worlds had to offer the opportunity for compelling storytelling. Planets would be rife with people, creatures, and opportunities for exploration; even the most iconic worlds from the films would have to be packed with surprises. "We have this giant canvas of known world we can use," says Senior World Designer Shawn Ketcherside. "But what's awesome is that, given the time frame of the game, we can take those expectations and expand on them. We can give fans something new and exciting."

FROM TEXT TO VISUALS

Final concept for the Valley of the Dark Lords on Korriban by Arnie Jorgensen.

Concept work for the worlds of TOR began in 2007, starting with Korriban and Tython. For many video games, writers provide a rough idea of a world or an area and the artists create it. Maybe there's going to be an ice area, a volcano, and a water area because the writers think it would be cool to have some variety. But that may be the extent of the guidance that art gets. With the story-driven nature of TOR, the writers conceptualized the planets and wrote detailed design documents for each world. The documents were passed along first to art director Jeff Dobson and then to the concept artists, who, led by senior concept artist and LucasArts alumni Clint Young (*Star Wars*: Republic Commando, *Star Wars*: Bounty Hunter, KotOR), would envision the worlds. This began a complex pipeline in which the writers and artists constantly gave feedback to one another. It was a true collaborative process in which story informed art, and all art served the story. Although not many people will notice the small choices that the artists have made in support of the story, they are huge parts of what makes the worlds feel real.

Before a single detail was added, the artists did a series of color studies, as many as twelve to fifteen in a day. "These are just straight up color. You're not actually painting anything," explains Young. "You're finding that palette. If it's an evil location, say, Korriban, you know that it's going to be bathed in warms." Once the color studies were done and compared against other planets to make sure the various locations had unique feels, the bare bones of the worlds began to appear in the paintings. Is the world soft and inviting? Cold and hard? "You start to introduce the edges, the silhouettes, and then from there you plan out the locations," says Young.

From there, the artists created more detailed sketches until deciding on a final set. For familiar worlds, such as Hoth and Tatooine, it was important not only to hit all the right notes, but also to give people something new. "You want to give people the dream," says lead world artist Andrew Collins, who also worked on *Star Wars*: Galaxies. "They want to be on Tatooine, so you really don't do a lot to pull it away from Tatooine; what you have to do is add to it." Dobson adds, "I love taking the real *Star Wars* stuff and nailing it, because if I'm feeling it, then I know that our fans are. The people who are even bigger *Star Wars* fans than me are really going to appreciate the work that was done to put them on these worlds."

Creating a new world meant an artist had more freedom, but less reference material. "If we want to add a color that's going to make the planet stand out or if we want to add combinations of colors or fog, then we can do that," says Collins. More freedom, however, often meant spending more time to make sure the place felt authentically *Star Wars*. The artists were clear that you can't just slap a random sci-fi planet into the galaxy and expect it to fit. There are rules to *Star Wars*. "It's simplicity turned on its ear," says Young. "If it's an ice setting, then it's an entire world that's ice. If it's a swamp world, it's the entire planet that's a swamp. In the later movies, it's a city planet." Architecturally speaking, those rules translate to rough buildings with smooth, rounded tops. Small widgets and unfinished details to the machinery. Exposed pipes and blinking lights. "It's blue collar. It's truckers in space," says Young. "It's a lived-in environment that you've never been to before, but looks like places that you've seen all your life."

Completed concepts were sent back for a gut check to the writers, who often embellished them with new fiction to explain interesting ideas or features. Designing Dromund Kaas, Dobson felt the world lacked something, and he remembers thinking, *What can we do to make this planet look cooler?* The concept team's answer was a series of majestic lightning towers that catch and pull electricity from the planet's perpetual storms. Then the writing team made those towers into Dromund Kaas's primary power producers and wrote a quest wherein the towers are under threat.

The last part of every world—and the one that players would probably think about the least—was the skybox. Skyboxes serve to ground the location, create the atmosphere, and determine the lighting. "Nar Shaddaa is dark. It's underlit. Tatooine—it's blue skies with wispy, high stratosphere clouds. It's like doing a matte painting," Young explains. "You're giving a sense of depth. It's like building a Broadway set. You want the audience to believe there is a sky there, that there's location beyond my location." When the illusion was complete and it looked as if there was always something to explore just over the next hill, a planet was ready to go in the game.

Paint-over of in-game screenshot to show new direction on Taris by Ryan Dening.

Hutta concept by Clint Young.

Sith blood pool by Zach Hall.

When the Sith Empire returned from a thousand years of exile to rebuild its holdings, its very first target was the sacred planet of Korriban. Home to the tombs of many of history's most powerful Sith, Korriban had been a place of pilgrimage and learning for the Empire, and it would become so again.

Korriban was the most popular world from the Knights of the Old Republic series and the first planet to make the cut for *Star Wars:* The Old Republic. In late 2007, it was also the first to be conceptualized. During these first explorations, the art team developed a detail-oriented approach to the planets and honed the final environmental art style. Arnie Jorgensen's first paintings of the sunset behind the pyramid of the Sith Academy, symbolizing the Emperor gazing down on his Empire, convinced the art team that the planets needed strong, specific lighting. This meant no time of day changes or other fluctuations. Each world would have a distinct look: Korriban's would be a fiery sky over red sand. Each world would also have its own themes encapsulated in every design decision. Korriban is about domination, and so everything that represents the Sith is monumental and imposing. Statues that are not of Sith Lords are supplicants, with spines bent as they literally support the weight of the Empire on their backs.

Korriban tombs by Clint Young.

Library in the Sith Academy by Clint Young.

Clint Young's original concept paintings of Tython.

Clint Young's original concept paintings of Tython.

The homeworld of the Jedi Order, where the Force was first discovered, Tython is situated in the Deep Core, its unbridled wildness far removed from the tranquility the Order enjoyed on its previous homeworld, Ossus.

Concepting for Tython began in late 2007, but the team developed it nearly from scratch because no visual reference for it existed. The artists imbued the planet—where the Force was discovered and the first Jedi developed--with a foreboding blend of awe and fear. "It's like being by yourself in an ancient forest," says Erickson. "Everything is gorgeous, but there's that feeling of power that underlies the whole thing that can be a little unnerving."

Tython was the second planet the design team worked on, and the world's rugged topography of thick forests skewed by protruding bluffs and promontories proved tough for the artists to capture. "Korriban was just rocks and sand, so there wasn't a lot to it. There weren't a lot of challenges there," says Jeff Dobson. "Tython is a complex, lush, populated environment, and it had us tackling a lot of problems that we didn't have to solve before." The designers tried creating the world's mountains by pulling up terrain and, when that didn't work, by placing huge rocks on the planet surface. Then they started over. "We kept throwing different people at it to see if somebody could stumble onto a new way to do it; we spent a lot of time figuring out what the heck we were doing," says Dobson.

Flesh Raider concept by Arnie Jorgensen.

The battle for Ord Mantell by Clint Young.

Separatist volcano base by Clint Young.

The government on Ord Mantell is corrupt in every way imaginable. As the Great War came to a close, a number of returning Ord Mantell veterans decided that instead of putting their weapons away they were going to throw the bums out forcibly. What started as meetings and vague threats became an open rebellion and then erupted into a separatist movement against the Republic.

Preproduction work on Ord Mantell—a world for which little visual reference existed—began in early 2008. "Everyone knows that Han went to Ord Mantell and had trouble with some bounty hunters," Daniel Erickson says. "In fact, there's quite a bit of stuff in the comics and in the books, but you don't see much of it." To expand upon the visual, artists for TOR developed a series of islands with magnificent cliff faces and an endless sea. Then they scarred all of it with bombs and littered it with the remains of broken transports, crashed ships, and mine fields. In the background, enormous turbolaser batteries shoot down enemy transports while bombers streak overhead. "We wanted somewhere you could feel what war does," says Erickson. "It was important that you could walk through the rubble and find small points of beauty. Think to yourself, This planet must have been gorgeous once . . ."

Hutt palace cantina by Clint Young.

CHOWBASO BUNKY DUNKO: HUTTA

While the city-moon of Nar Shaddaa is the central base of operations for the Hutt Cartel, Nal Hutta is the homeworld for the powerful species. Often called *Hutta*, the planet is everything one would expect of a retirement paradise for rich Hutts: disgusting, polluted, self-indulgent, and depraved.

"We've never seen Hutta fully realized like it is in our game, so designing it from the ground up was really fun," says Clint Young. But *fun* doesn't mean *easy*, as Daniel Erickson remembers. "Hutta is supposed to be awful. Gross. Horrific," he says. "But of course you don't want to put players somewhere they won't enjoy themselves. So you have to imply awful, while making it inviting." The art team did this by mixing the iconic shapes from Jabba's palace and skiff in *Return of the Jedi* with a color palette that was occasionally described as "pea soup and barf." Huge pipes of sewage were added to show the Hutt's complete disregard for the environment, and massive factories were inserted in the otherwise endless swamp to give players some recognizable geography. Inside the buildings, Hutt excess meant cantinas where parties never stop, in plush surroundings just out of range from those pipes pushing out toxic sludge. All of the elements came together to create a quasi-realistic swamp whose details are just a bit cartoonish—a terrible place to live, but a whole lot of fun to visit.

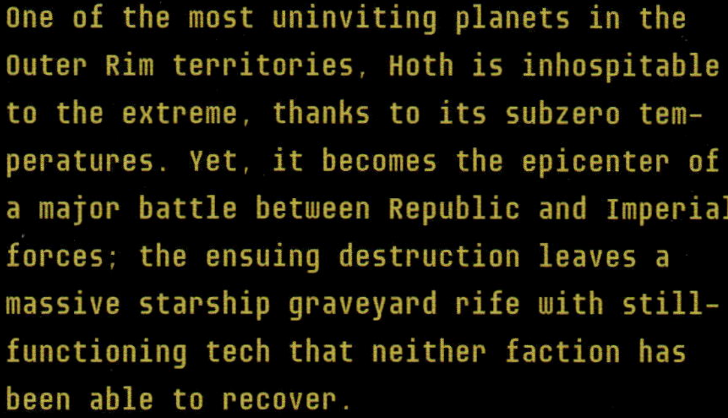

Early Hoth concepts by Clint Young.

TUNDRA PLANET: HOTH

One of the most uninviting planets in the Outer Rim territories, Hoth is inhospitable to the extreme, thanks to its subzero temperatures. Yet, it becomes the epicenter of a major battle between Republic and Imperial forces; the ensuing destruction leaves a massive starship graveyard rife with still-functioning tech that neither faction has been able to recover.

Hoth has been depicted in a voluminous number of Expanded Universe sources—after its introduction in *The Empire Strikes Back*—as a flat, desolate world punctuated by snow banks and the occasional cave. TOR's writers and artists felt this depiction limited the planet's appeal to gamers, who would be spending hours on it. Given creative license to introduce new areas and geological elements, the writers added a starship graveyard, an Imperial base, and a giant pirate outpost. They also included volcanic mountains jutting out of the snow, based on established lore that Hoth is heated internally by volcanic rock—which also creates enough heat for vegetable matter to grow in caves and feed native tauntauns.

Given that playable areas were separated from one another by huge expanses of snow, the artists added visual flairs to break up the monotonous geography. "We did different things, such as balancing expansive areas of snow with ice structures, cliffs, and ice lakes, more mountain ranges in certain areas and less in others; we broke up those spaces so they didn't feel boring," says Jeff Dobson.

Early Hoth concepts by Clint Young.

Ice volcano by Clint Young.

Paint-over of in-game screenshot by Clint Young.

A hardscrabble world of rock and sand baking in the oppressive heat of twin suns, Tatooine is of interest to both the Empire and the Republic. The Imperial military has forced aside underworld figures to assert its dominance, while in Anchorhead the Republic is secretly establishing an outpost of its spy network.

The birthplace of the Skywalker legacy, Tatooine is the cradle for the entire *Star Wars* saga. Since the planet has appeared in five of the six live-action films, the art teams had a wealth of reference available when they set out to concept Tatooine in early 2008. Still, the artists had to re-create accurately what was in the movies while retaining the game's stylized aesthetic. "We're less realistic, less gritty, and we have a little bit more of a fantastical approach to our coloring and our design," explains Andrew Collins.

The artists also needed to give *Star Wars* fans something new, so they consulted Expanded Universe literature for ways to embellish the planet. During their research, they stumbled upon a little-known geological tidbit: The planet is able to sustain small pockets of fertile vegetation and oases. The trick was to visualize them in a way that retained the planet's bleak ambience. "We don't want a paradise if Tatooine is this desolate, dry, desert planet," says Collins.

Desert camp by Clint Young.

Nexus Room Cantina by Ryan Dening.

IMPERIAL THRONEWORLD: DROMUND KAAS

A remote marsh-world of eternal night and electrical storms caused by ancient dark-side rituals, Dromund Kaas was rediscovered by the Sith after the Great Hyperspace War. The Emperor viewed the world's inhospitable climate and isolation as the ideal capital for his burgeoning Empire.

Created as a major set piece for *Star Wars:* Jedi Knight Mysteries of the Sith (1998), Dromund Kaas features a verdant landscape punctuated with ancient temples and catacombs, implying historical ties to dark-side practitioners. When the writing team decided to establish the world as the seat of the Sith Empire, concept artist Clint Young approached it with the aftermath of the Sith seizure in mind. "Dromund Kaas in its day was probably very beautiful, lush and green, lots of color, until the Empire moved in and set up fortifications," he says. "It's almost as if the planet starts to rebel against the people who are actually living on it, so you get beautiful areas that have become corrupted. It's dark, it's foreboding, lots of kinetic energy, and not all of it good."

Young collected reference imagery of canyons and forests and twisted their shapes while refraining from completely subjugating the world to corruption. "You want to show the beauty that's left over from the world."

Color study for Imperial Citadel by Ryan Dening.

Broken city by Clint Young

THE BROKEN WORLD: TARIS

Life in the ruins by Clint Young.

Acid lake by Clint Young.

Once a highly sophisticated Outer Rim ecumenopolis rivaling Coruscant, Taris was bombarded by Sith Lord Darth Malak in one of the most savage war crimes in galactic annals. The Republic, returning to Taris three hundred years later in an attempt to rebuild and use it as a symbol of recovery against Sith destruction, has discovered a world of horrors.

Clint Young's Taris concepts depict a postapocalyptic wreck of a world strewn with the remains of towering skyscrapers that once dominated the landscape. Although imaginatively conceived, they confounded the design team tasked with creating the vision in three dimensions. "Taris was *nasty,*" says Andrew Collins, who recalls the first time he saw Young's designs in early 2008. "It was really cool, but we had to make it look as if you were at the bottom of a city that had been destroyed years ago. You know when kids get a giant tub of LEGO bricks and they dump it on the floor? That's what Taris looked like."

The planet artists literally redesigned Taris from the ground up by scaling back much of the devastation and very slowly blocking out the planet piece by piece. "We decided that players would be on the ground, where everything fell, and we treated Taris as a basement where pylons that once held up the city were," says Collins. By creating those pylons, a plate structure, and debris, the team completed the first pass of Taris. "It was good and would work for anybody who wasn't a hardcore KotOR fan, but something was missing," says Dobson.

Taris was set aside while work continued on other planets. When it came time to optimize Taris, Collins, Dobson, and Jorgensen decided to start from scratch. They shut down everything else they were working on and spent the next six weeks creating new textures and terrains. Jorgensen hung vines from the broken plates, implying that local flora had overtaken the world once it fell victim to disrepair. Meanwhile, Collins added architectural variety and more layers to the wreckage. "It needed to feel as if it was once a city," he says. "What you're walking on should feel like the fallen plates that existed before, and we added more trash piles, areas of fallen debris." Finally, the artists built huge, twisted beams to represent the ruined superstructure of the city that had once stood there. "I was like, 'That's it! There it is!'" recalls Dobson. "It finally felt as if this was a giant city that had been knocked to the ground."

Alderaan mountains by Clint Young.

Killik cave by Arnie Jorgensen.

he majestic feel of Alderaan's
n fields and frosty mountaintops
tles of the nobility. Soaring
impossible architecture give a
et still medieval feel to the
t farther from civilized spaces
the Republic's oldest ruins—and
rther, the Killik. A sentient
ies explored in a series of EU
Killik create the large mounds
their hives, adding yet another
y detail to a beautiful but
dscape.

Various Coruscant signs by Arnie Jorgensen.

Coruscant underworld mood shot by Clint Young.

Various Coruscant signs by Arnie Jorgensen.

JEWEL OF THE GALAXY: CORUSCANT

The broken heart of the Republic, Coruscant is a world of endless city. Social status is closely tied to altitude, with the richest and most powerful citizens living up high. Living conditions become more impoverished toward the surface, where huge, ancient machines run the vast infrastructure. Poor citizens and alien residents swarm into market and industrial areas, set up barricades, and declare themselves squatters; in deeper levels, gangs rule over fiefdoms like warlords.

In late 2008, Coruscant was the first city-world the team tried to design, and it presented a unique challenge for which they were not prepared. Usually, artists shape a planet by lifting and manipulating natural terrain digitally and then placing separately produced objects on top of that terrain. This wasn't going to work for Coruscant, which has zero natural terrain. The entire world was going to have to be built from objects—but separately rendered objects are far more expensive than terrain per digital square kilometer. The answer was to develop a small set of reusable pieces with wildly different looks on their sides: For instance, one piece might have a factory wall on one side, housing on the other, and a shop piece on the third (the same technique was used with physical pieces by Industrial Light & Magic to construct the Death Star for the first film). By manipulating and moving these pieces around, the art team could build a city with all the variety required and still run at the demanded frame rate.

The Prophecy, a central Voss religious structure, by Ryan Dening.

Nightmare Lands by Clint Young.

A planet of more questions than answers, Voss was discovered by accident shortly after the Treaty of Coruscant. Its rocky plateaus, unspoiled peaks, and verdant forests were at first thought to be inhabited only by a reptilian species, the Gormak. When a second, much less populous species makes itself known on Voss, however, the galaxy is thrown into turmoil.

Voss head concepts by Arnie Jorgensen.

Sometimes a new planet in TOR starts with a visual concept—Belsavis's deep jungle-filled fissures or Dromund Kaas's perpetual lightning storms. Sometimes a planet is the best backdrop for a story essential to the game's overall themes—the corrupt but Republic-friendly government of Ord Mantell or the Imperial efforts to drain the Republic's resources with the unwinnable war on Hoth. Voss was created in 2008 to give a home to the Mystics, a group of Force users with no knowledge of the Force as the Jedi and Sith knew it. The Mystics had been considered as a potential character class early in the game's production, but were rejected as too alien and requiring too much explanation. Voss gave them their own world.

The Voss Mystics are Force users of incredible power. Both the Sith and Jedi agree on this one point. The Voss do not. The Voss know nothing of the Force and are uninterested in outsider opinions. The writers asked the artists to put the Voss somewhere isolated and overwhelmed. In response, the art team designed a single city, Voss-ka, high atop a defensible mountain and surrounded by the encroachment of the Gormak. Ruins of abandoned Voss holdings were placed to show the path of the Gormak's advance; deeper in the wilds, ancient ruins of Jedi and Sith origins spoke to rumors that this was not the first discovery of Voss. And beyond that, darker things still. "The single biggest win on Voss were the Nightmare Lands," says Daniel Erickson. "A part of Voss was supposed to be corrupt and spreading—the artists created something far moodier than I'd ever expected." Trees twist and impossibly sharp rocks push through the ground to create an area where everyone who has come before has gone mad—and players must, of course, go in and explore.

GORMAK

Tech-savvy but pre–space flight—and extremely hostile—the reptilian Gormak make their debut in The Old Republic. "The Gormak have the typical *Star Wars* tech: the greebles, the pipes, all the stuff that humans use, but they're covered in this armorlike skin," says Clint Young. Whereas most things in *Star Wars* look machine-made, Gormak tech by contrast feels molded and precision cut. "It's got this high gloss, as if everything has been wiped down with oil," says Young. "It's like an alien race living in the Renaissance era and manufacturing a lot of their equipment. You can still see the guts, the machinery underneath exposed. You can tell that they've taken a little bit of who they are and put that into their technology."

Gormak head concepts by Diego Almazan.

Belsavis nature studies by Clint Young.

Belsavis nature studies by Clint Young.

Anomid concept heads by Paul Adam.

lent of Alcatraz. "You're sitting inside one of these craters and there are glaciers all around you, blocking out the sun because they're so huge," says Young. "It's like, *Where am I going to go if I escape? Nowhere!*" Young incorporated immense ice fissures, deep inside of which thrived teeming areas of wilderness. "I thought that maybe the inner planet core was heating these little pockets, causing these giant glaciers to melt; in doing so, it exposed all of this under-earth to the sunlight, causing lush vegetative forests to grow inside the glaciers."

ing; I can imagine that it's like living in Miami, where you're sweating all the time, but everything has this lift to it."

Jungle steam crater by Clint Young.

City traffic by Clint Young.

The Promenade by Arnie Jorgensen.

"The Smuggler's Moon." "The Vertical City." "Little Coruscant." Nar Shaddaa is the capital of the Hutt Cartel's criminal empire and the galaxy's dirtiest city in every sense of the word. While the Sith Empire openly participates in the slave trade and even the Republic has been known to turn a blind eye to smugglers loyal to its cause, only Nar Shaddaa plays host to every crime, every depravity. Cybernetics, genetic modification, organ smuggling, designer spice—if it's a horrible idea that the galaxy could easily do without but is worth credits, Nar Shaddaa can supply it.

Nar Shaddaa began being visualized in early 2009. "We've seen Nar Shaddaa several times in LucasArts titles, and we've taken that image and the concept and just pushed it," says Young. "It's Vegas, but not the 'Bring your family to Vegas.' It's the old Vegas. It's nasty. But on the flip side, it's got that 'Wow!' factor." Huge skyscrapers are covered in neon; holographic signs of fifty-foot-tall dancing girls lure patrons to the red light district; and the night sky is lit with advertisements and the celebration of a thousand parties. Hutt pleasure barges ply the sky lanes along with off-world shuttles filled with wide-eyed tourists hoping for a bit of adventure.

Signage found throughout Nar Shaddaa by Clint Young.

Droid factory on the plains by Clint Young.

Republic fighters streak over a ruined landscape in this color study by Clint Young.

Caught in the throes of the Great War between the Republic and the Sith, Balmorra still strains under regular skirmishes between the Sith Empire and pockets of resistance fighters who battle against Imperial control over the planet's droid and weapons manufacturing plants.

Balmorra was a testing ground for various effects and environmental details that the art team would later spread throughout the game's war-torn worlds. The art team worked out how to do flight paths for background ships flying in the atmosphere, create laser missile shields that can be bombarded in real time, and have giant turbolaser batteries constantly firing at enemy targets. Each of these elements added atmosphere to the game's war zones, but each had its own challenges. The missile shield required a half-dozen iterations before the team found something that worked and that didn't slow down the game when missiles began to fall.

Concept for bombing runs on Balmorra by Clint Young.

Quesh Venom running on the planet surface by Clint Young

Tatooine's heat and Hoth's freezing cold can't compete with Quesh's constant 90 percent humidity, made up of toxic gases. Originally harvested for the production of low-grade poisons, Quesh gas is also the basis of experiments that revealed powerful adrenals. Though the Republic controlled Quesh, the Hutt Cartel controlled the technology, so a secret deal was struck. When the Empire discovered the scheme, it forced the Hutt Cartel to join in an invasion of the Republic-controlled world.

"Quesh is not a pleasant place," says Daniel Erickson. "You need a gas mask or a series of injections to survive there. It's a dead world." Creating the unusually wet and active wasteland of Quesh meant understanding what the world had been before. The present Quesh atmosphere was once subterranean gas, the bulk of which is still underground. When the gas first leaked to the surface after a quake, centuries before the game's time period, it killed all of the flora and much of the fauna. So among the mines and production equipment are skeletons of massive creatures that once walked the planet's surface.

Dangers of a Quesh gas mine by Arnie Jorgensen.

THE CRYSTAL WORLD: ILUM

Defense of the mines by Clint Young.

On the surface of the ice-world Ilum sit two aging Jedi temples of meditation, several ruins of early exploration, and little else. Below the surface, however, run the richest veins of Adegan crystals in the galaxy. These crystals are used to make lightsabers, and the Empire has also been doing research into their ability to bend light for cloaking devices for their ships. Conflict is inevitable.

First seen in the 2003–2005 Genndy Tartakovsky *Clone Wars* cartoons, Ilum has some obvious similarities to Hoth. Where Hoth has low, rolling snowbanks and large, flat spaces, Ilum is jagged with cliffs and precarious drops. When first deciding to do the world, the writers thought that difference would be enough to set it apart visually. The artists disagreed. Their vision of Hoth had developed beyond the ice plains of the movies; Ilum was also going to need something else. "The artists came to me and said, 'We really want to do Ilum at night,'" remembers Erickson. "The argument was that at night in Antarctica you can see forever. It's so cold, so crisp, that it's almost as if you're in space." One look at the stark, almost alien landscape created by the environmental art team and everyone was on board. Ilum would be a night world of huge battles, where you could see blaster bolts from vast distances.

Commercial district by Arnie Jorgensen.

Destruction of Axial Park by Clint Young

If Coruscant is the Republic's heart and Alderaan its soul, then Corellia is its sense of independence and adventure. A heavily populated urban center that resisted the overbuilding typified by Coruscant and Nar Shaddaa, Corellia is famous for its ship making and its live-and-let-live attitude, which attracts misfits, rebels, and artists from across the galaxy. This was before a group of Corellia's ministers decided to turn their world over to the Empire in exchange for permanent ruling positions. The Corellian people rose up in massive, organized rebellion when their freedom was threatened.

Coronet City on Corellia was a city very different from the urban designs that TOR artists had developed for Coruscant and Nar Shaddaa. Coronet City had surprising green spaces, wide roads, and a fantastic public transit system. "Part of the early idea for Corellia was to show the conflict between who the Corellians pretend they are and how they live," says Erickson. "These are supposedly the most self-interested people in the Republic, and they've set aside beautiful parks, have blocked development of skylines, and have a wonderfully functioning city." At least they did until the Empire invaded. First, the world designers laid out the city as it originally looked in all its beauty; then the artists went in and destroyed it, taking the war damage of Balmorra and Ord Mantell and multiplying it. This is TOR's hottest war zone, with whole areas under constant bombardment—but also with curiously pristine places where blaster fire can be heard just around the corner.

ndustrial zone by Clint Young.

In 1977, *Star Wars* amazed audiences with jaw-dropping moments that they previously could not have imagined—but perhaps none was quite as *strange* as the Mos Eisley cantina scene. It was one thing to hurl so many strange creatures at viewers at once; it was another to show them conspiring in dark alcoves, hunched over drinks, taking drags from space hookahs, and looking the other way when someone's arm is lopped off in a bar fight. Aliens are quintessentially *Star Wars,* and they're all over the place in The Old Republic.

A k'lor'slug larva design by Zach Hall.

Concept artist Diego Almazan developed many of The Old Republic's aliens and monsters. "There's a lot of low contrast in *Star Wars*," Almazan says. "You generally have a lot of creatures that are one or two colors with different kinds of noise going on within each of those spaces." By *noise* Almazan means *details*, all of which depend on the species. For a lizard species, it could be scales; for a frog, it would be warts and wrinkly skin. "We wanted different ways to add interest without necessarily jumping out at you, saying, 'Look at this great detail!' But if you're in close, you have something cool to look at. If the alien is far away, its look is still readable as one big, coherent shape."

Creating aliens and creatures that are fantastic yet anchored in reality and biologically sound is the key. "If you're in the desert, you expect to see lizards with big, ruffled scales that are meant to dissipate heat, but which stay low to the ground and keep under the rocks," Almazan continues. "If a creature's got to fly, it's got to be able to work; but as soon as you shrink it down, say, to the insect world, the rules are very different, because gravity is different."

"Part of the fun is looking up as many pictures as you can of weird, gangly frogs and spiders, and even monkeys and newts and lizards and whatever you can find that's relevant," Almazan adds. "My favorite is always looking up sea creatures, because you pull up the weirdest, gnarliest stuff. You say, 'This isn't from Earth,' but sure enough, it's deep within the ocean. I want to make the weirdest thing you've ever seen."

Various experimental creature sketches by Paul Adam.

The creatures of Voss look unique in the galaxy. This one was created by Sperasoft.

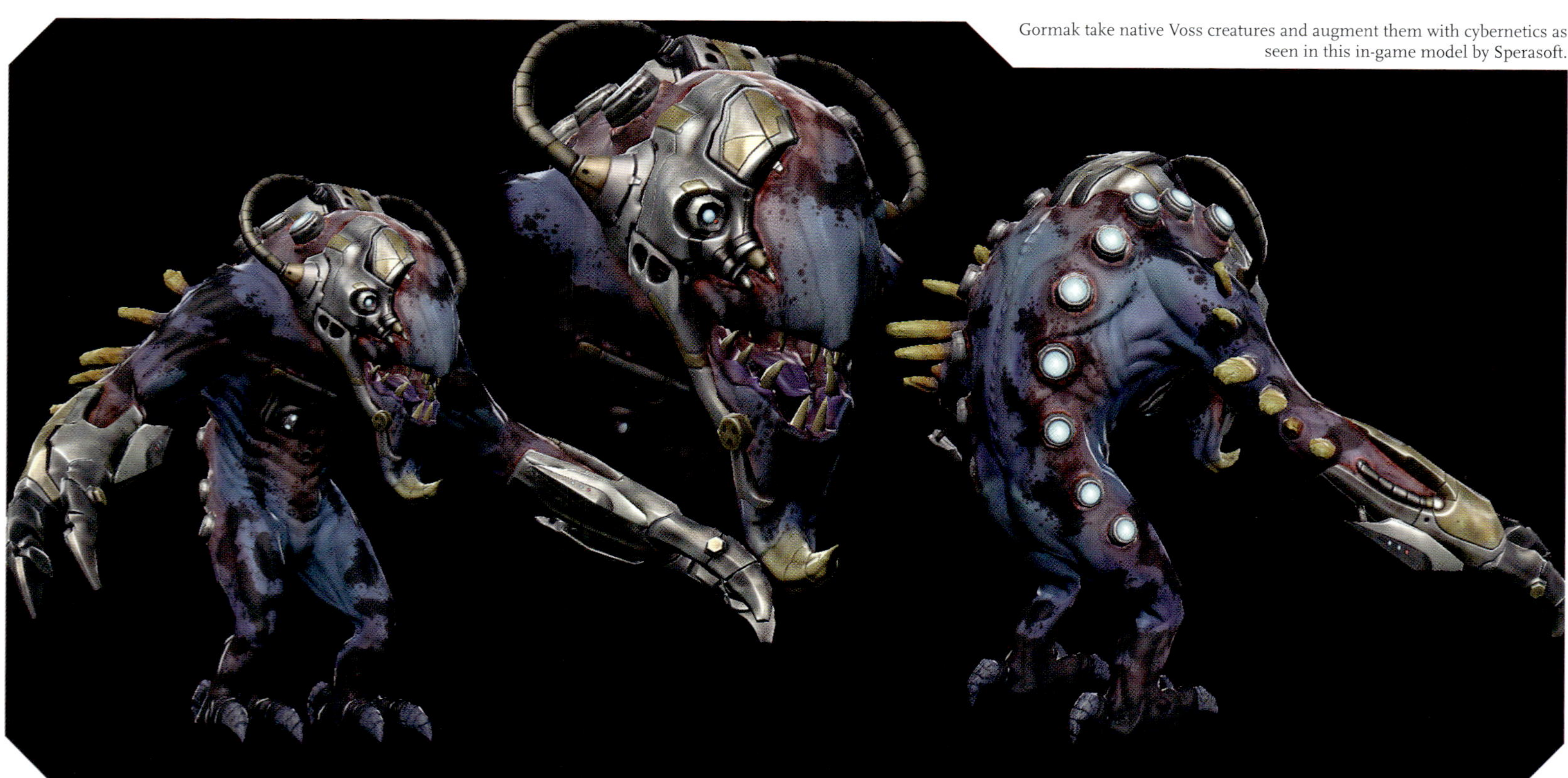

Gormak take native Voss creatures and augment them with cybernetics as seen in this in-game model by Sperasoft.

Early Terentatek concept by Arnie Jorgensen.

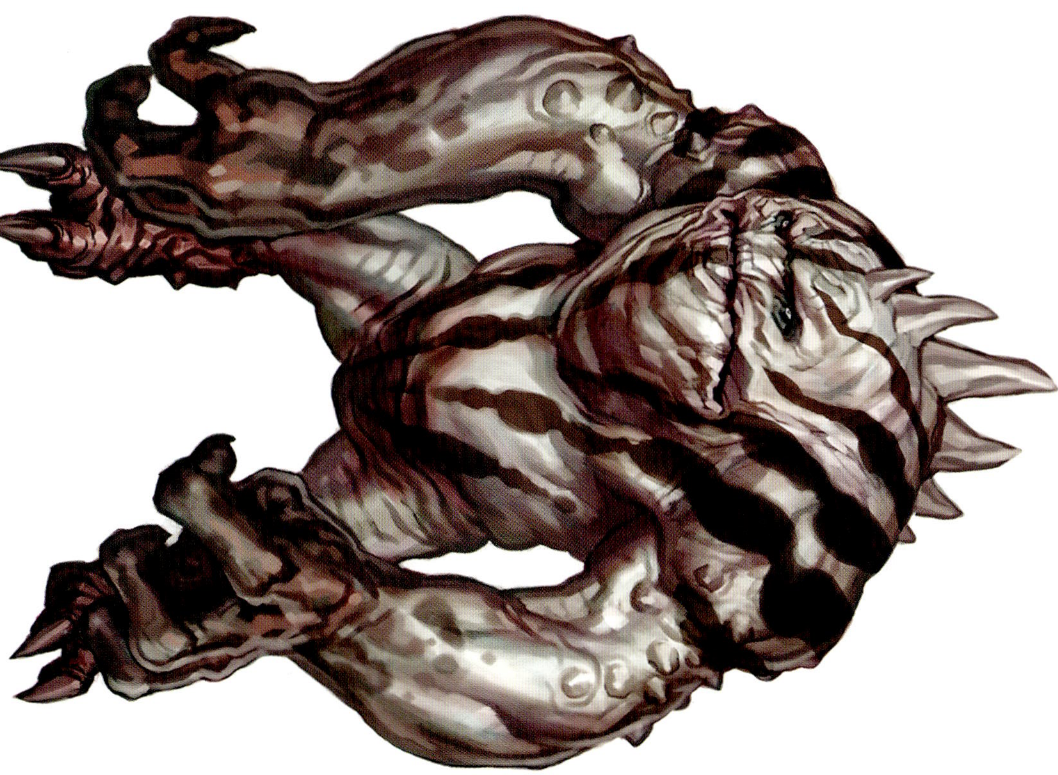

Kintan by Zach Hall.

MOBILE LEVELS: PLAYER STARSHIPS

From Luke Skywalker's X-wing, Darth Vader's customized TIE fighter, and the *Millennium Falcon* to the *Ebon Hawk* in KotOR, starships in *Star Wars* are practically characters themselves. Massive in scale and representing freedom and adventure, player ships in TOR act as the bases of operations from which players explore the galaxy.

At first, the plan was to create three ships: one for the Republic, one for the Empire, and one for the underworld classes (the Smuggler and Bounty Hunter). As concept art developed, however, it became clear that Trooper fans were going to want military vehicles, that putting an Imperial Agent in a Sith ship wouldn't make sense, and that Bounty Hunters didn't want clunky cargo freighters. The solution? Each class type would get its own incredibly detailed, customizable ship. In the end, the player ship count would double to six with the Jedi and Sith classes sharing two between them. "It's cool to give the player a ship, but now I have to do that times six. It's still awesome, though. It's something I will fight through. I will drag myself through the muck so that players can have ships," says Brad Prince.

Ryan Dening experimented with dozens of different looks for the Bounty Hunter ship.

Various hull ideas for the final Bounty Hunter ship design by Ryan Dening.

Senior concept artist Ryan Dening led the designing of TOR's player ships. Dening and the other artists first sketched rough exterior shots, experimenting with scale and imagining the uses for each craft depending on class. Once a particular concept was chosen for further exploration, the artists would develop landing gear, external engines, ramps, and other details, all the while ensuring everything had a *Star Wars* spirit. "Some of the ships feel very cobbled together," says Dening. "You can see all the heavy plates, even on the Imperial craft. A TIE fighter has little bitty things all over it. Some of the shapes are very strong and simple. It doesn't always have to make aerodynamic sense with a *Star Wars* ship. You have a lot of leeway."

Final Bounty Hunter ship design and exterior details by Ryan Dening.

After locking down the exteriors, artists moved on to interiors, a long, laborious process. First, they determined size requirements: How many rooms and for what purpose? All starcraft needed medical facilities, some would need meeting rooms, and others would require cargo areas. Each would have a central hub—or lounge—which established the feel for the entire ship and the player class to which it belonged. Next, the artists created a series of isometric cross-sections, mapping out the interiors of each ship. This determined the layouts of all of the rooms—props, lighting, mood—and how all of it worked together. "Oh, man, starships. That's the greatest thing in the world," says Brad Prince. "Giving the player a starship is so *Star Wars*. Having your own mobile house is just awesome. It's as if you're in the movie."

When the artists began concepting the Sith player ship, they had already developed an Imperial starship style: geometric, with lots of angles and smooth, flat surfaces. Since the Sith class ship would be factory issue, it couldn't be a deviation from that aesthetic so much as a dark twist on it. "You were given this ship by the Empire, so it wasn't some evil ship concocted out of the dark side of the Force or something," says Dening. Dening tried different designs until homing in on one that was symmetrical and that amalgamated iconic shapes from the films' spacecraft. "It's as if a TIE interceptor and the *Millennium Falcon* made out and had a baby," says Dening. "It's got the docking room at the back, and it's got some turrets on the sides, but it also has these big triangular wings."

Sith player ship cockpit by Ryan Dening.

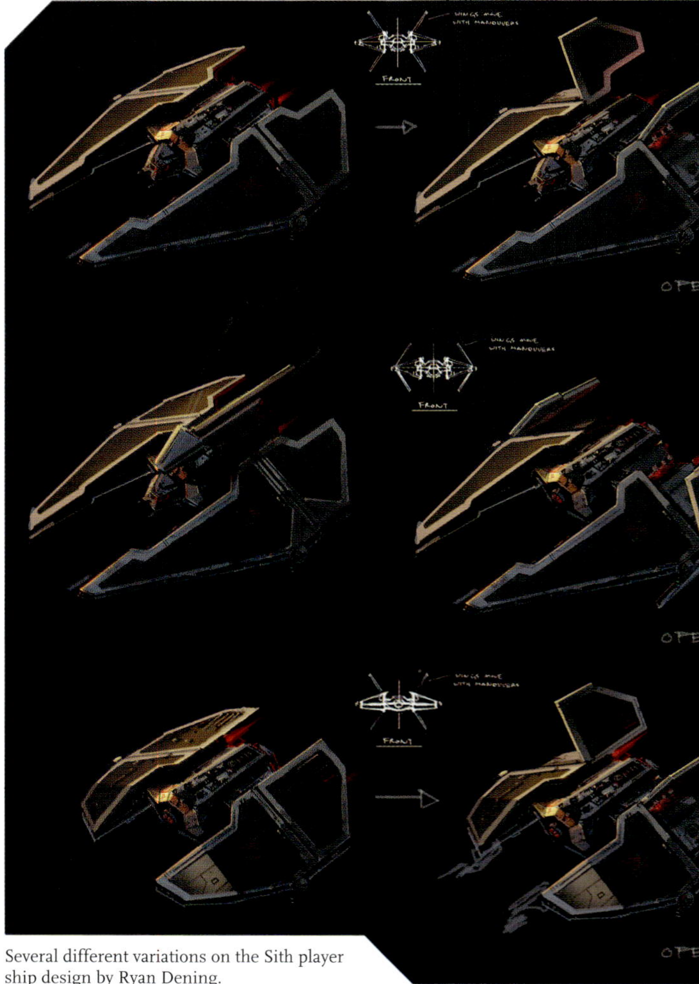

Several different variations on the Sith player ship design by Ryan Dening.

Jedi player ship color variations by Diego Almazan and Ryan Dening.

Diego Almazan provided the original sketches for the Jedi class ship, based loosely on the *Consular*-class space cruiser seen in the opening scene of *The Phantom Menace*. Dening brought in details from *A New Hope*'s Rebel Blockade Runner and the hammer-headed *Endar Spire* from KotOR, turned sideways. Given that the Jedi Consular class would use the ship, the designers' first instinct was to create something innocuous and diplomatic, but early explorations in that direction were rejected. "It just didn't feel like something you would want to fly around space and shoot stuff up with," says Dening. "So we put a lot of muscle on it. Now it feels a lot bigger than the other ones, even though it's not; but it has some real presence."

Clint Young based the Smuggler-class player ship on the *Millennium Falcon*, another step toward providing players with a vicarious Han Solo experience. "It seemed fitting," says Dening. "If you wanted to play that fantasy, you'd want to fly in the *Falcon*." Young also worked in aspects of the *Ebon Hawk*, such as side-mounted cargo containers and two large, rear engines, as well as turrets and a protruding cockpit to give it a "trucker in space" look.

Senior concept artist Paul Adam drew upon the B-wing starfighter when designing the Trooper ship, adding a large gun, a wing, and a cockpit to the sides. "It's somewhere between a Humvee and a big, armored personnel carrier," says Dening, who brought in heavy beams, huge shock absorbers, and a command center to give the Trooper ship a cramped—though functional—feel. "It feels as if the whole floor is suspended, in case they get hit. I would assume they do get hit a lot, since it's the Trooper ship. It feels as if it's a battle machine or a submarine." For the Bounty Hunter class ship, Dening used the shape of *Slave I* as the basis. Like *Slave I*, the Bounty Hunter ship features engines and gun placements coming off the sides, and it changes its orientation when going into flight mode. "It feels like a bird of prey," says Dening.

Trooper player ship design and details by Paul Adam.

Smuggler class ship by Clint Young.

Smuggler class ship interior by Ryan Dening.

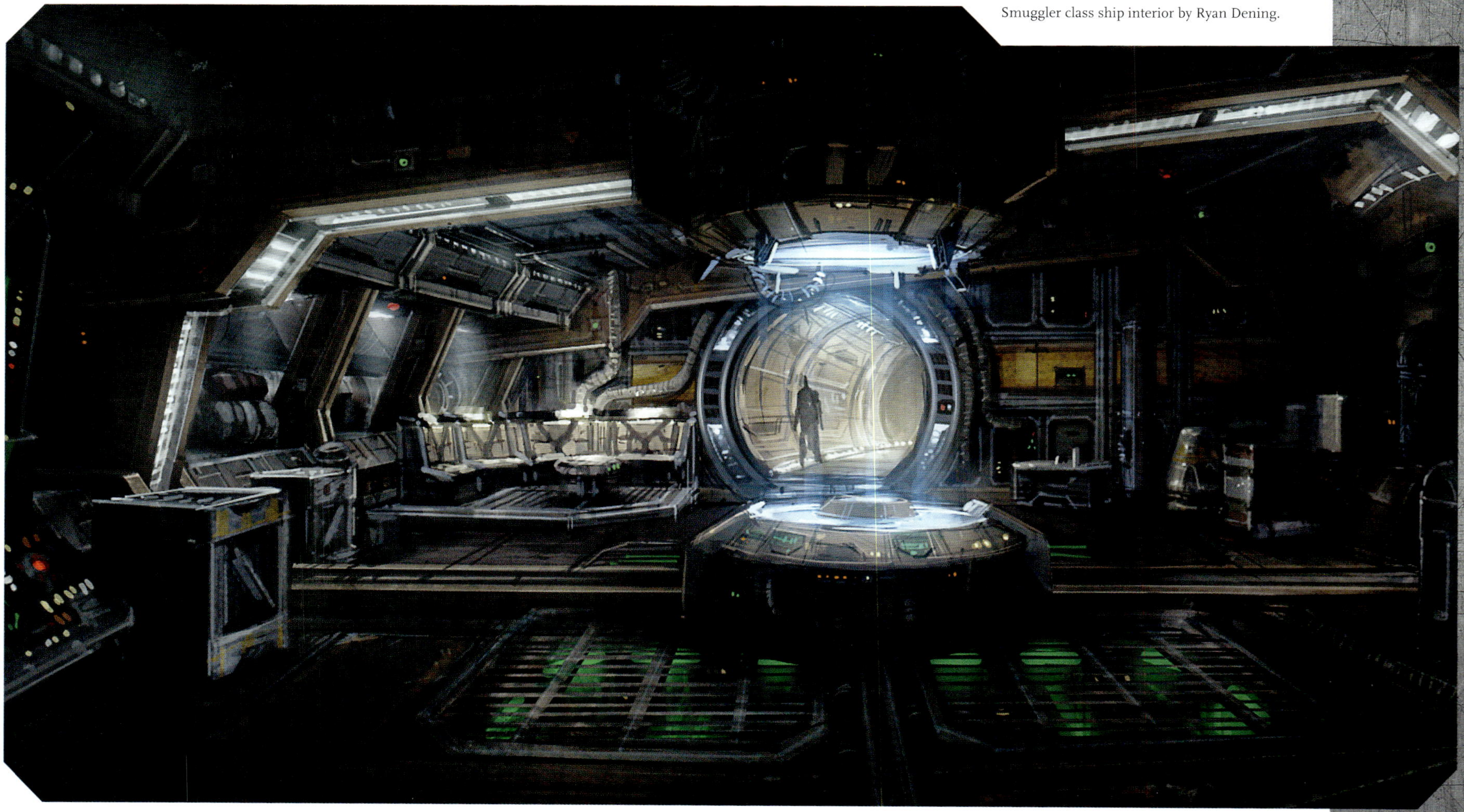

The Imperial Agent ship, like the class it was designed for, was one of the most challenging to create. Even though the ship belongs to the Imperial class, its undercover nature meant that it couldn't be Imperial issue. Also, no clear analog existed for a spy ship in the *Star Wars* films. Looking for a solution, Dening married the Naboo royal starship from *The Phantom Menace* with an Aston Martin and the SR-71 Blackbird reconnaissance craft, resulting in one of the slickest ships ever seen in the Expanded Universe. "It feels very much like a high-tech supership," says Dening. "It just has a character totally different from all the other ships."

Imperial Agent class ship by Ryan Dening.

Imperial Agent class ship interior by Ryan Dening.

IMPERIAL AND REPUBLIC HARDWARE

Republic capital ship concept by Ryan Dening.

Dening's starcraft designs didn't stop with player-class ships. *Star Wars* being, well, *Star Wars*, spaceships are the primary mode of transportation and are therefore omnipresent. It was important that The Old Republic's frigates, fighters, and capital ships have a signature look and feel, while remaining consistent to the look of their respective faction's line.

By harking back to Imperial ships of the original trilogy, Dening's Imperial ship designs connect players with *Star Wars* but also hint at things to come. "It's a pretty classic Imperial look," says Dening. "Our capital ship is almost a Star Destroyer. We're really playing up that fantasy of the Empire, and figure maybe the new Empire, in the original trilogy, harks back to the glory days and they're trying to bring that back."

Dening gave Republic-issue craft a slightly jury-rigged look with warm colors and rounded surfaces to contrast the more angular and cold Imperial-issue ones. "It's a little bit more slapped together."

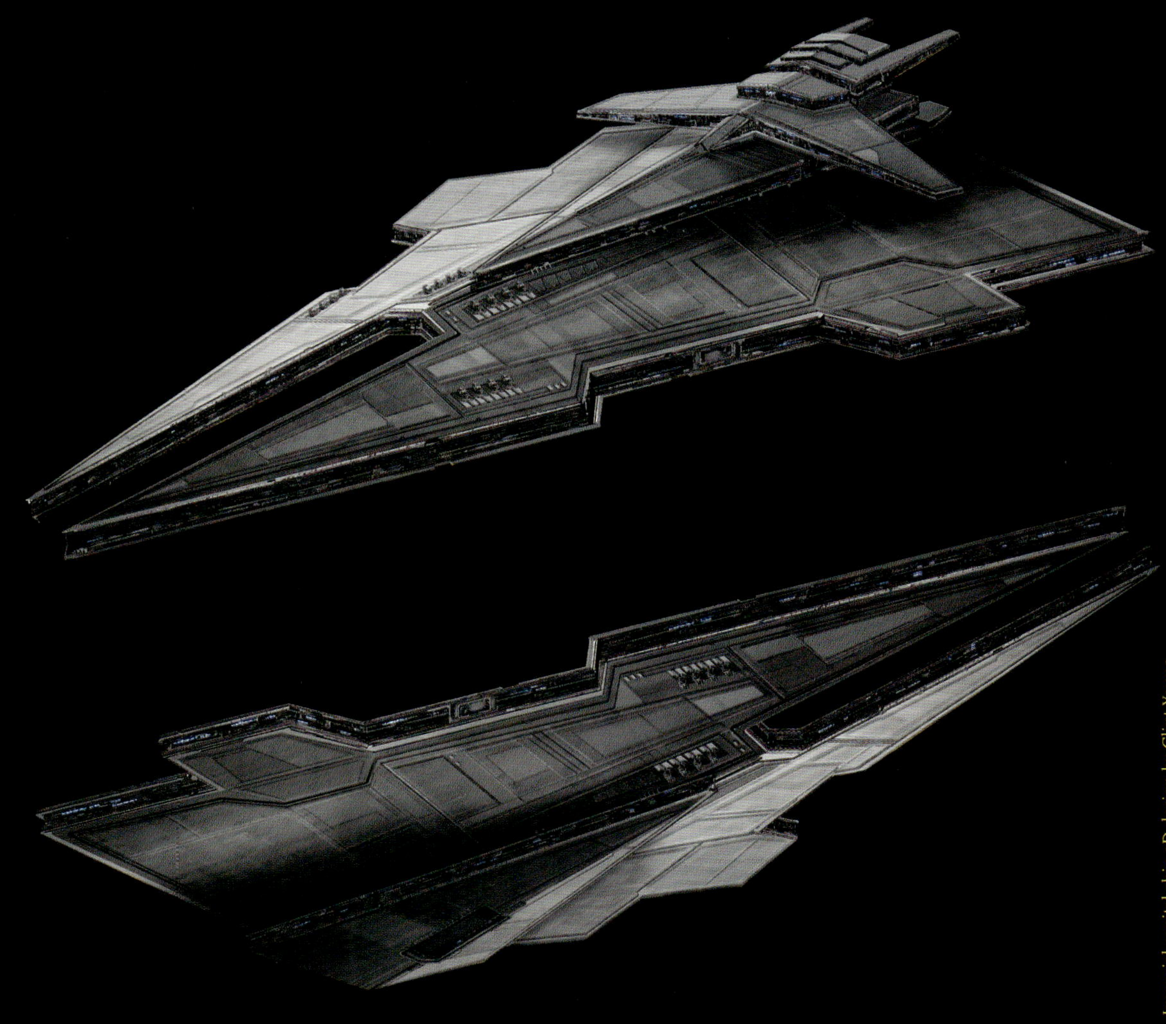

Imperial capital ship 3D design by Clint Young.

SPACE COMBAT

Space game mission design concept by Casper Konefal.

Always on the team's wish list, but an intimidating undertaking, was the space game. Providing customizable ships for the various classes was complicated enough, since it would enable players to carry their level with them wherever they went. Taking those same ships into space to do battle would require a completely different game, seamlessly integrated with the primary experience. "As it turned out, the game engine was quite versatile. This allowed us to deliver an eagerly anticipated—yet novel—experience to The Old Republic player," says Kevin Barrett, executive producer in charge of the BioWare team that crafted the space game.

With a love of games such as X-Wing and Rogue Squadron, the team designed a space shooter, heavy on cinematic graphics, light on complicated starship controls. *Star Wars* is about fantasy action and dogfights after all, not the complex maneuvering and tactics of *Star Trek*. Built against beautiful backdrops, the game would be fast and exciting. "When we watched the space battle sequences from all six *Star Wars* films for the umpteenth time, one of the things that stood out—beyond the fiery explosions and blaster crossfire—was that in *Star Wars*, space is never empty blackness," says Elliot Christian, the space game's lead artist. "We've gone to great lengths to create an exciting variety of space terrain. We then composed those elements to create a close shave here, a beautiful vista there; and so everywhere we provide star pilots with a dynamic and cinematic experience."

Space game look and feel concept by Clint Young.

Space game mood shot by Clint Young.

Space game mission design concept shot by Clint Young.

An early color study for the space game by Ryan Dening.

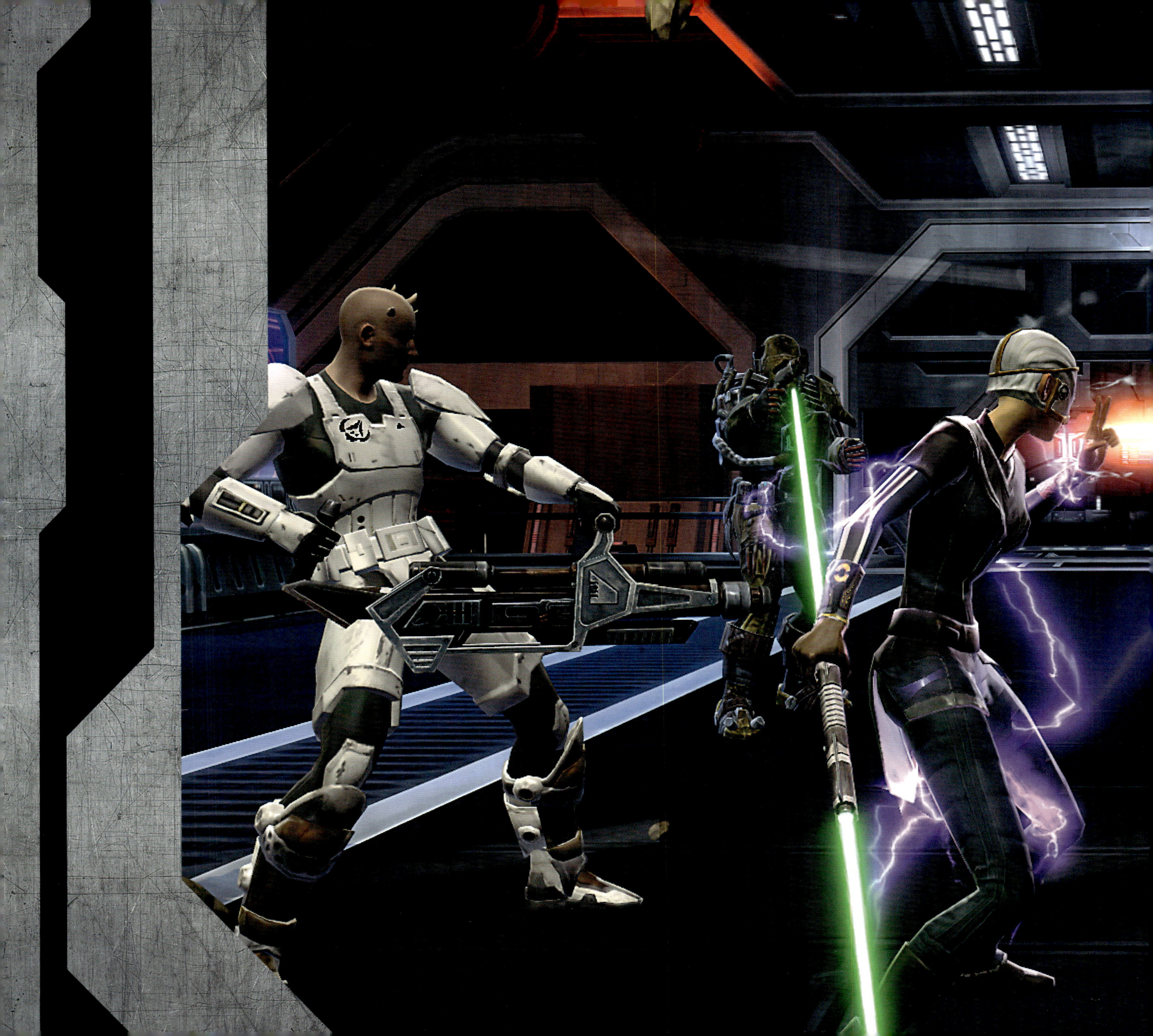

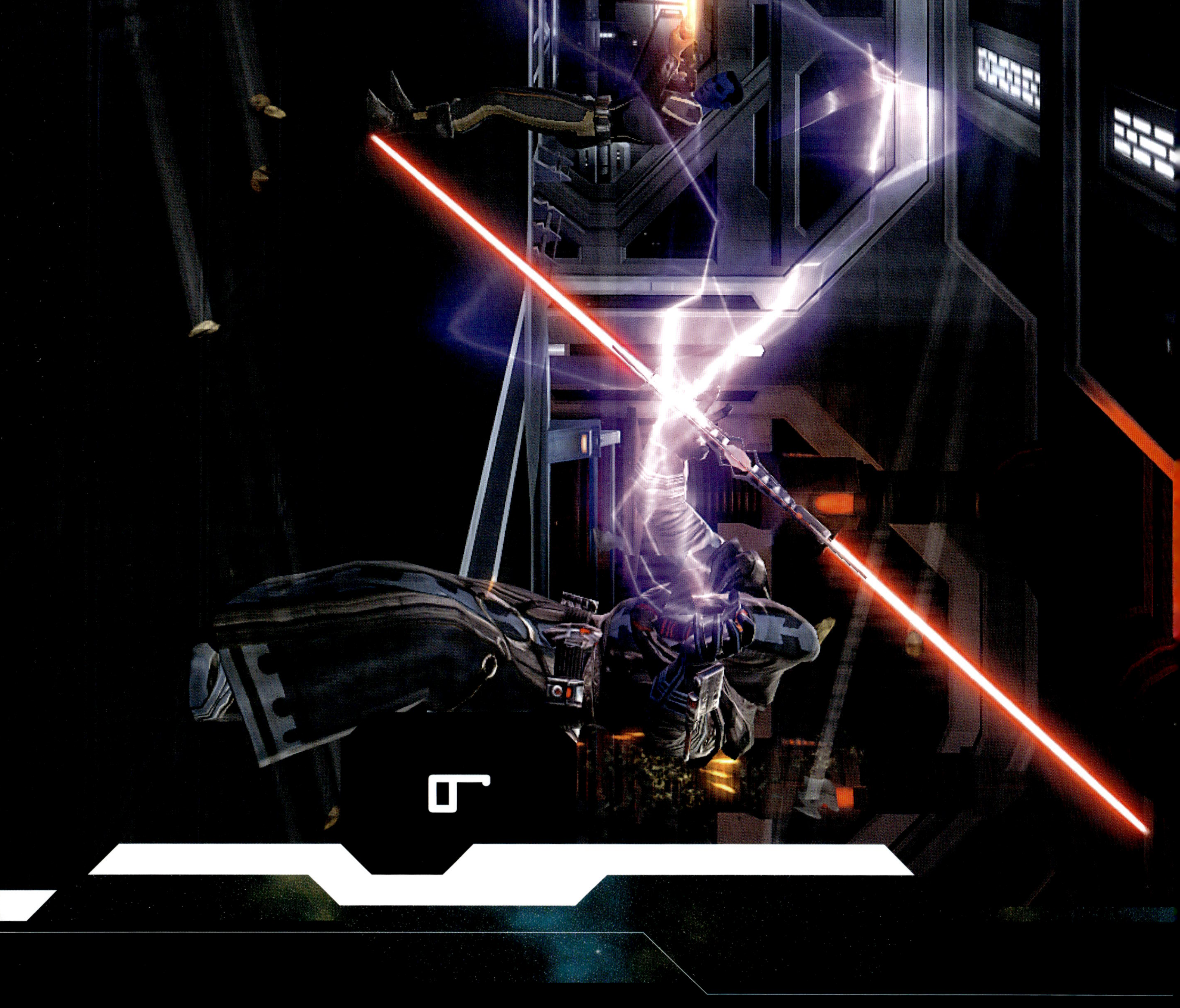

6

MULTIPLAYER ADVENTURES: INNOVATIONS AND OBSTACLES

AT SOME POINT, I'M SURE EVERYONE READING THIS BOOK HAS HAD AN IDEA FOR A GAME OR GAME FEATURE. THINKING OF NEW IDEAS IS THE FUN PART. BUT, MAKING THOSE IDEAS COME TO LIFE IN A WAY THAT PLAYERS WILL ENJOY IS AN IMMENSELY COMPLEX ENDEAVOR. CREATING SOMETHING THAT'S INNOVATIVE YET FAMILIAR IS A DIFFICULT AND REWARDING CHALLENGE."

CORY BUTLER

BIOWARE
PRODUCER

PAGE 140–141:
Players clash in the Alderaan Warzone.

Although the developers put a great deal of emphasis during early development on the class storylines that would take players through the eight unique experiences, it was clear that the focus of the game would be on community. The Old Republic progresses certain MMORPG staples, such as crafting and combat, while adapting systems from single-player BioWare games, such as story and companion characters, to a massively multiplayer setting. Having to integrate single-player elements and multiplayer action required the developers to surmount plenty of technical barriers and to bring forth a number of game-play innovations.

The first Flashpoint for Imperial players involves assaulting an enemy ship.

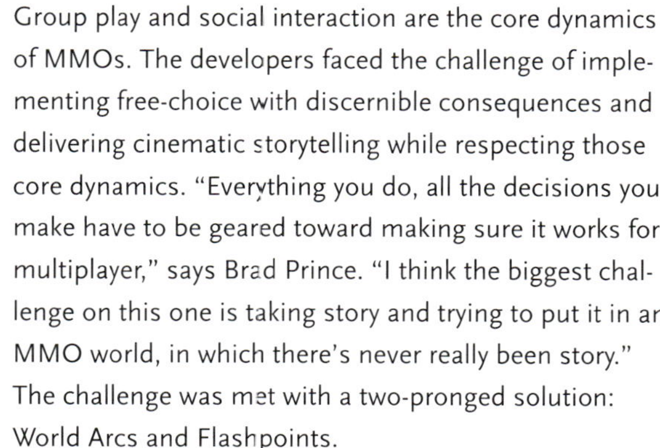

BRING ALONG YOUR FRIENDS

Group play and social interaction are the core dynamics of MMOs. The developers faced the challenge of implementing free-choice with discernible consequences and delivering cinematic storytelling while respecting those core dynamics. "Everything you do, all the decisions you make have to be geared toward making sure it works for multiplayer," says Brad Prince. "I think the biggest challenge on this one is taking story and trying to put it in an MMO world, in which there's never really been story." The challenge was met with a two-pronged solution: World Arcs and Flashpoints.

One characteristic of classic MMOs is that almost everything that takes place in public space is a one-off quest or a short chain of events. But that allows for little or no character development, no plot twists or surprises. Events but no story. The designers created World Arcs to change that. A World Arc is a chain of up to ten quests that span an entire planet and may take twenty hours or more to complete. Each quest builds off the last and players must make key decisions at several points. World Arc adventures are very close to playing a traditional RPG, but with your friends.

Traditionally in MMOs, the only attempts at building a story were in instanced dungeons. "I'm a huge fan of doing multiplayer stuff . . . and for me, dungeon crawling is the ultimate thing to do," says Prince. "I love them so much that it's an aspect that I wanted to take to the next level." Hence, Flashpoints: handcrafted cinematic adventures for one person or a party closed off from public zones. Flashpoints not only offer visceral cinematic thrills, but also provide players with highly scripted, character-changing decision points. Players get to make huge choices that change the entire plot and resonate across the galaxy. They have the chance to destroy capital ships, invade space stations, and experience complex storylines that rival the best of single-player games. "That's hard to do in a public space, so you put that into an instance zone," says Prince. "You can have timed explosions, you can have scripted events, you can have all the cool stuff that really seems like you're playing a movie."

COMPANION CHARACTERS

In TOR, the strongest relationships that players form are with their companions, the AI-controlled characters that fight alongside them and join them on their travels. Companions give players advice and have distinct feelings and personalities. They are guides to the intricate, choice-driven stories, and each comes with personal storyline as well. And they roll into *Star Wars* perfectly. "If you think about it, the movies are about companions," says Rob Cowles. "Han and Chewie, Luke and R2-D2—when you have a *Star Wars* adventure, you probably would adventure with a companion."

Companion characters have long been a staple of BioWare games, and the developers pushed that concept to enhance single- and multiplayer elements of The Old Republic. With the addition of companion characters, players always have a wide array of support when playing alone. In multiplayer, companions can be used to customize groups and ensure that different roles are being filled. "Companions can fit a variety of different roles," says Jake Neri. "Companions mean you never have to go on any adventure alone."

Smuggler and companion, Bowdaar.

Jedi Knight and companion character, T7.

Sith Warrior companions Vette and Pierce team up to do some crafting.

Jedi Consular and Trandoshan companion, Qyzen.

Companions would also be instrumental for crafting—creating goods such as weaponry and armor—a vital component in any MMO. How crafting systems work varies from game to game. Some systems reward dedicated crafters who establish businesses, craft for others, and build reputations and skills that other players seek out in the game world. Other systems cater to casual crafters who prefer to spend their time adventuring. Principal lead systems designer Damion Schubert (Meridian 59, Ultima Online) oversaw virtually every non-combat play mechanic in the game. His team designed a system that could support both the casual and the dedicated crafter.

But the *Star Wars* setting, which lends itself to heroic fantasy, posed a unique problem. "There are some people who are not in love with the idea that a Sith warrior would ever craft a robe or bunny slippers," Schubert explains. The solution? Players in TOR can direct companions to craft for them. "The crafting system that we've devised is ultimately more about you being the Donald Trump of your little economic sector. You are the guy setting up the business, doing the research, trying to get all the pieces in place, but you have your companion characters do all the work for you so that you don't have to get your hands dirty. Hopefully this will actually elevate the player and make him feel more like a big shot, more in charge."

Associate lead designer Emmanuel Lusinchi took things a step further when he proposed a mission system with which players could send companions on noncrafting activities, such as running missions while the player is out adventuring. "The idea of directing your companions on your ship to craft or to run missions while you're absent is the idea that these guys are out there doing something, progressing your cause, whether it's your crafting skills or getting more quests for you, while you're gone," says Schubert. "It creates the sense that all of these guys still have value to you."

A Jedi padawan faces her first fight on Tython.

A Republic Trooper fires a heavy assault cannon.

A Sith Inquisitor uses Force lightning to deadly effect.

RUN 'N' GUN

Combat would prove to be trickier for the designers to visualize, let along implement. Creative director James Ohlen was adamant about not innovating in the nuts and bolts of combat. MMOs are run by a central game server, and that server has to judge which actions taken by players are legal and has to determine the outcome of any conflict. Akin to running every play in a football game through the referee before the fans get to see the ball thrown, this aspect is complicated and slow and lends itself to a deliberate, low-action style of fighting. Despite these restrictions, however, Ohlen was just as adamant that it couldn't *look* like traditional MMO combat. Often MMO fighting involves avatars facing enemies, each playing its separate, very basic animations. Combat in TOR needed to look and feel like *Star Wars*. That meant lightsabers impacting each other, characters running while they shot blasters, Force lightning, jetpacks, and other spectacular effects.

Above all, the combat needed to feel *heroic*. In traditional MMOs, players fight against small numbers of enemies, something that wouldn't resonate with the *Star Wars* tradition of underdogs contending against overwhelming odds. It also tended to not be until max level that players really felt powerful. Ohlen wanted to change that for Star Wars fans who would be expecting incredible things from the first moment. Especially from the Jedi, who should never seem anything but amazing.

However, in order to deliver a fun game in which all classes would be balanced in PvP and bring value to group situations, the designers needed to make certain concessions. "The classic debate that you see on the boards all the time is people arguing that a lightsaber should always kill things in one blow and should always sever people's heads in one swing," says Schubert. "It is very easy for you to get too wrapped up inside the fiction and say, 'Oh, this can never happen inside *Star Wars*.' But our top priority is always about making a fun game, building a strong community . . . and that is where you have to figure out how to make the concessions and balance them against really bringing the spirit of *Star Wars*."

As good as this all sounded on paper, implementing it in a massively multiplayer environment was another story. Combat in MMOs consists of living people—often with disproportionately sized avatars—fighting in unpredictable ways. Choreographing ranged and melee combat for convincing attacks, defenses, and counters was a joint effort by the animation and programming teams. The animators began by choreographing wireframe skeletons of two opponents to create baseline animations. From there, actions were blended in a process known as additive animation, enabling the programmers to determine and control character body positions and match their movements toward larger or smaller opponents. Eight character classes and divergent body types all had to be cross-referenced so that skeletons of a particular size could be choreographed against larger or smaller skeletons.

COVER SYSTEM

To give the Smuggler and Imperial Agent classes an edge against the Force-using and infantry classes, designers developed a cover system that enabled those classes to exchange fire while safe behind protective objects. A first in MMO history, the cover system was an entirely new concept for the designers and programmers. "It was a challenge from the animation standpoint in terms of how the characters were going to move and what that meant for the server, the client, and everybody else in the world," says lead animator Mark How. "It surprises me when you're seeing different kinds of fighting and you get this guy who rolls behind this cover object, firing and shooting. I expect that in all sorts of games, but I'd never seen that type of combat happening in an MMO. It intrigues me and it baffles me in terms of how [well] it turned out."

A firefight on Hutta.

An MMO is easily gaming's most challenging technical puzzle. Besides being immense in scope, the game runs on a live server that must remain online and consistent.

BioWare has created several game engines in the past, even licensing those engines to other developers. When it came time to create its first MMO, however, the company quickly made the decision to license rather than build. "We were following the 80/20 rule: If we could license an engine that did 80 percent of what we want, we could build the other 20 percent," says former technical director Bill Dalton. "And the usual caveat is that that will take 80 percent of the time. The real play for us was to find a working engine so that we could actually have the content people going early, because this BioWare game had some new ideas about what an MMO could be." With the need to prototype early and often, the team had to have a platform it could start using right away, even if it meant spending the next few years gutting and replacing parts in a process Dalton described as "like replacing the engine on a motorcycle while you're riding it seventy miles per hour down a busy freeway."

One stipulation for any licensed tech was that the team would get full source code, which means that the licensing company allows the team access to every secret of how the program works. Often a licensing agreement keeps parts of the source code out of the hands of the game developers in the interest of trade secrets. But because an MMO is a live service, it was important that the team be able to instantly fix a problem instead of calling the original developer and waiting for a response. HeroEngine was one of the few engines the team looked at that actually met all of their criteria: They could get the source code for it, and it worked. Simutronics, the engine's creator, was in the process of building an MMO with the engine, so it had been field-tested in a limited way.

At the time, it was not a proven release engine, however. While Simutronics had significant experience building MMOs, the HeroEngine had no track record, no released games to point at and compare what would work or not, be simple or difficult. Still, it allowed the team to, literally, start building on day one.

The engine a game is built on is one of the biggest determining factors to the speed and ease of development.

TECHNICAL HURDLES

The game development itself took place in a live environment, similar to how the game would finally ship, which added an extra layer of difficulty. "The advantage of developing in a live environment is that everybody on the team is exercising the code, the logic, and the content simultaneously, all the time. And so any problems that crop up are evident immediately and presumably can be fixed fairly readily," explains Dalton. "The cost of the live edit environment is that if we break the editing system, it's worse, because now people can't enter content at all until those systems are fixed. So it's very hard to maintain and expensive to develop, but when it works, it works very well."

The combination of a live environment and a largely untested engine that was constantly being reworked resulted in a number of memorably painful moments during the game's creation. These included a bug whereby one person could lock out all of the content the other hundred or so developers needed to work on. This was especially painful, as it tended to happen when the person was checking work in for the day and going home, so that person couldn't easily be found to revert it. Other fun bugs included hundreds of characters turning invisible seemingly at random, creatures with textures that would explode off their bodies to fill the screen, and an infamous Destroy the World button.

"There used to be a command in HeroEngine that would instantly flatten all the terrain on a planet," says Erickson. "And it wasn't a complicated command. It was one of the function keys and Ctrl or something. And *boom!* Whole world is flattened—all work destroyed." To make it even more interesting, the game environment did not, at the time, have a working Undo feature. One was quickly added.

CONTENDING WITH SCALING

From a development standpoint, the primary difference between an MMO and a single-player game is that developers have a very different sort of control over the world scene in an MMO. "With a single-player game you can drive the player through a series of interactions and you're leading them to some reward, or putting them through a lot of torture to give them a payoff at the end," says Bill Dalton. But MMOs don't work that way; they're emergent. "MMOs tend to follow the open-world, sandbox approach. Here's a giant world that you can go explore and find some fun that way, but we're not going to lead you through it in any meaningful way."

More to the point, other people come in and explore, too. Players interact with one another, often in ways that even the designers don't foresee. The inability to control what players do or the kinds of events they create in the game space is compounded by scaling. Massive amounts of data about character and environment positions and movement are constantly communicated back and forth between clients (the players' machines) and servers (which supply the data) over a network. As the player base grows, high volumes of player traffic can strain networks and create performance problems for the servers.

Warzones were one of the first places to really test the technical skills of the team.

"If everybody wants to come to my apartment in my virtual world, there's nothing anybody can do to prevent that, and that may put two hundred to three hundred characters on-screen at once," Dalton says. "How well does your game client hold up to that? How well does your server hold up to that? Those are the kind of rivet-popping points in the development of a game that differ from a single-player game, where you have a lot more control over that."

According to Dalton, the first line of defense against this was game design. "We want the game design to encourage people to cover the terrain as evenly as possible, and so they have to usually pick up a quest at point A, run out into the world, and then return to point A. But we don't want people to just sit at point A for indefinite periods of time without having any kind of fun, because if everybody starts doing that, we hit that scenario."

The programmers also put strategic measures in place: reducing the frequency of updates sent between client and server, reducing the fidelity of those updates, and even reducing visibility to avatars who are in the immediate vicinity. "Practically speaking we might just show you the nearest thirty-five player characters and just not even give you updates on anyone else. But those are last-ditch, worst-case scenarios, where we have to prevent the server from grinding to a halt," says Dalton. "We have to prevent the client from being completely unplayable."

Boss fights are a mechanism for getting groups together.
There is safety in numbers.

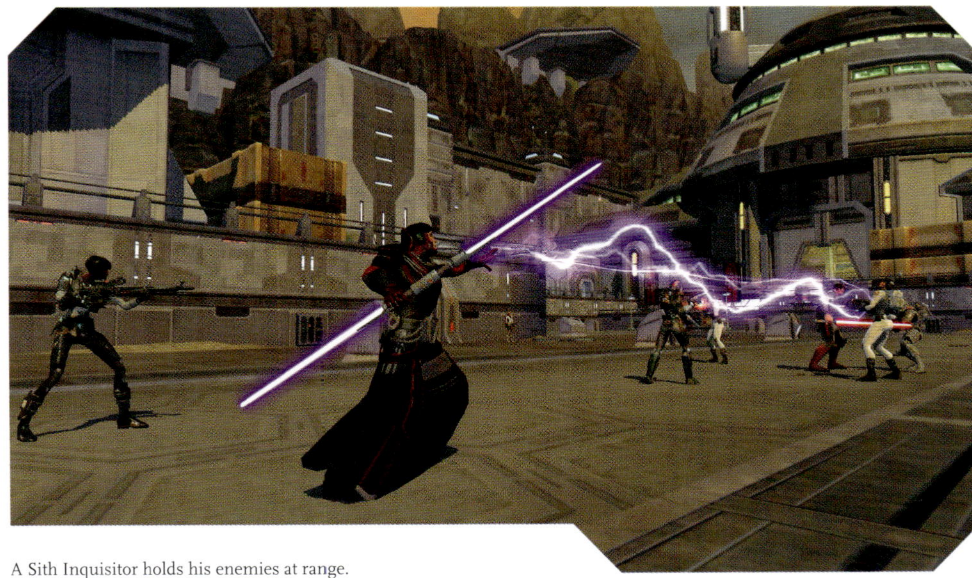

A Sith Inquisitor holds his enemies at range.

A pair of Republic Troopers take point in a firefight.

"My earliest *Star Wars* memory is the movies, and wanting to be and do what Luke was doing or Han Solo was doing," says lead animator Mark How. A childhood spent playacting scenes from the *Star Wars* trilogy, How says, gave him the creative impetus for breathing life into The Old Republic. "You wanted to have Force powers, and so you would try and push stuff around or slam into your buddy and do different kinds of things like that. It's a direct relation to animation, to combat, and how the characters should move."

To help sell the game's sci-fi fantasy milieu to players, characters needed to display complex, realistic physical movements. Motion capturing actors would yield better, faster results. All actors need direction. So the animators studied all six *Star Wars* movies and filmed themselves having firefights and lightsaber battles with Nerf guns and prop weapons—often in office hallways, to the chagrin of their coworkers. This helped inspire the performances and ensure that all animations would resonate with what fans see on-screen. "You can have a guy swinging a sword or a guy shooting a gun in many other games, but that doesn't make it *Star Wars*. It doesn't make it a lightsaber or an actual blaster," says How.

The Force-using classes required additional animation to convey convincingly their superhuman speed, leaps, and flips. Fastened to harnesses, stunt actors were motion captured while bouncing off trampolines and being hurled around the studio. The takes from those sessions would be fed into 3D Studio MAX, in which the animators animated on top of the footage to make characters jump higher and move faster, repositioning and reposing them as necessary. "You get human movement, which satisfies the base that we actually work from, and then you turn it into Force jumping, which is beyond what a regular person can do," says How.

Motion capture only gets the basic data.
Every animation needs to be hand touched.

Each animation is reviewed repeatedly before it's put in game.

Although a number of the designers had worked on previous BioWare games that featured dialogue cutscenes, many had come from MMO development backgrounds and didn't have experience creating the type of emotionally charged, cinematic conversations that The Old Republic required. This would fall to Lead Cinematic Designer Paul Marino and his team of specialists. Cumulatively the game's cutscenes would comprise hundreds of thousands of lines of dialogue, a daunting prospect from an animation and programming standpoint. "As soon as you go into a cinematic piece you have characters with their lips moving, that are animating; you have to have good camera angles," says Brad Prince. "You can't tell a BioWare story without those cinematics, and the scope on a game like this, it's mind-boggling." Manually creating every animation and positioning every camera would have been physically impossible; systems that could automate a bulk of it were required.

The cinematic designers used a customized version of a tool that had originally been developed for Mass Effect nicknamed RoboBrad (after Brad Prince) to assign simple animations for lip-synching, head and eye movement, glances, and gestures based on dialogue. These baseline animations gave the designers a foundation on which to iterate by hand. For camera placement, a system called RoboDirector automatically assigned cameras according to the dialogue and the number of people in a scene. Then the designers went in and fine-tuned those camera positions. "It's quite impressive to click a button to generate an entire conversation and then watch that conversation and say, 'Hmm, that's pretty darn good. It looks like somebody hand-touched that,'" says Prince. "You get some weirdness here and there, but it becomes a job of cleanup instead of starting from scratch."

Complex cinematic scenes are the hallmark of all of BioWare's games.

Originally, nonplayer characters were to be fully voiced, with the player-character dialogue conveyed by subtitles. This was itself an ambitious proposal considering that the genre's traditional method of storytelling relies on text boxes to represent dialogue. But the stakes increased after BioWare released Mass Effect, in which all characters—player included—were fully voiced. "James and I were testing an early build of Mass before any of our VO was recorded," says Daniel Erickson. "We looked at each other and knew player VO was going to be the future."

"At the time we realized, *OK, this is nuts!*" says Darragh O'Farrell, recalling the decision to go with full voice-over. Although O'Farrell's team had never worked on a single project the size of TOR, they had worked on the first two KotORs, the Monkey Island games, The Force Unleashed, and a bevy of other titles. They were accustomed to tackling large projects with aggressive deadlines. "Back-to-back, sometimes in the same year," says voice director Will Beckman. "We could handle it."

Casting

The writing team began delivering scripts in early 2008. They called for sixteen voice actors for the character classes (eight classes, male and female versions for each), almost three hundred thousand lines of nonplayer character dialogue, and thirty thousand lines of alien languages. "If you were to take all the voice-over from Mass Effect II, Mass Effect, Dragon Age, Knights of the Old Republic, and Baldur's Gate and combine them, it still wouldn't equal the amount of voice-over in *Star Wars:* The Old Republic," says Ohlen.

O'Farrell's team immediately rolled up their sleeves. Beckman cast his net wide, calling agents and conducting open casting calls in New York, Chicago, and London, attracting stage and screen talent. Convincing performances would be key in creating emotional bonds between players and their characters and companions. Finding the right actors was crucial. "You're not looking for someone who's going to change their voice and do ten different characters; you're looking for naturalistic performances that fit snuggly within those characters," says Beckman. Adding to the challenge, TOR players would be able to choose body types (from wiry to heavyset), alien species, and ethnicities. "You need people that fit those four very disparate models," says Beckman. "If you get someone too light and that person is supposed to inhabit a big, heavy guy or girl, it just doesn't feel right. The actors had to have voices plain enough to map all those varieties."

Casting the NPCs was its own organizational feat, as Beckman needed to map actors to thousands of roles across different planets, ensuring that players would feel immersed in a galaxy teeming with variety. No one wanted players to talk to an NPC, turn a corner, and talk to another one voiced by the same actor.

Recording

Recording sessions began in spring 2008 at more than ten studios throughout the world. Voice directors were present in every session—during which up to four hundred lines would be recorded—and sometimes BioWare's writers would call in to give further guidance. They had to help actors understand their roles in the context of the game's branching stories. "A lot of times you have people who have never played a game and aren't thinking in a nonlinear way," says Beckman. "Because the story is modular and can go in all these different directions, people will ask, 'Do I like this guy? Do I hate this guy?' Acting for a play is like expressionism. It's emotive and broad. Acting for a film is a realistic style of acting. Acting for a game is like cubism. You have to be able to take it all apart and see it all from many different angles. Then we have to make sure that it all fits back together again."

Cha skrunee da pat Sleemo!

Beyond the main dialogue, thirty thousand lines of alien-speak were recorded. To put this in perspective, thirty thousand lines of dialogue is more than the entire script for Mass Effect or both KotOR games combined. Just in aliens. "You're going to see tons and tons and tons of aliens in this game," says Kellogg. "We're trying to do as much as we can to build a robust system that gives a lot of variety, so that every time you talk to a Rodian, you're not hearing the same lines over and over again."

The scripts called for thirty-three new languages, most of which were created from scratch. Beckman started work in spring 2009, using real languages as bases—Finnish, for example, which he used as the template for Ortolan—and altering their phonemes to create something original. For Huttese, Beckman recorded actors reading the script, used effects to create a compressed version, and mixed the two to create a dense, warbling speech. "The game has a bit of an exaggerated art style, so the sound design is a little bit more exaggerated than what we hear in the movies," he says.

Some species voices were hybrids: voice files so heavily altered that they ultimately became sound effects. According to Expanded Universe lore, the Gand don't have vocal cords but speak by exhaling, causing their exoskeletons to vibrate audibly in an approximation of verbal language. "The question was, 'How are we going to do *that?*'" says Orion Kellogg. "Do we have a sound designer create sound effects or do we have an actor read lines and have a sound designer process that to make it sound more like a vibration language? It's iterative and takes a lot of trial and error to get it sounding right."

"When a game character is voiced, there is a sense of authenticity and significance associated with the player's interactions, which changes the perception of the game character," says Todd Davies, BioWare's lead audio designer. "You then either relate and empathize with the character or you choose not to. When is the last time you said to yourself, 'Wow I really care and relate to this text pop-up box'? Immersion, emotion, and connection are not effectively conveyed by a text pop-up box, no matter how fancy or elegant its border."

"In the past, typically, if an MMO shipped with two hours of music, you were pretty lucky," says LucasArts music supervisor and staff composer Jesse Harlin. In May 2009, Harlin tapped lead composer Mark Griskey (Knights of the Old Republic II) and Gordy Haab (*Indiana Jones* and the Staff of Kings) to help compose music for TOR. Four additional composers would join the effort, and together they would create more than six hours of new orchestral and cantina music. More than two hours of other music, culled from both KotOR games and John Williams's music from the films, would be judiciously added to provide subtle links to TOR's past and future. "We're taking pains to make sure that we're faithful to the films in that themes associated with particular planets, characters, and concepts are respected," says Harlin.

The bulk of recording took place at Skywalker Sound in California and in Bratislava, Slovakia, over a three-week period in December 2009, an extremely short time considering how much material was written. Musical themes in *Star Wars* traditionally have been identified with specific characters. But because players would be creating their own characters in TOR, the composers wrote themes for each class, ensuring that each had a signature instrument, style, or melodic sensibility. The Trooper class meant lots of brass and percussion instruments. Flutists used a wind-instrument technique called flutter-tonguing to create a buzzing, slightly ambiguous sound that gave the Agent a sense of mystery and uncertainty. Scraped piano, a technique in which a hand is scraped across piano strings, was used to give the Bounty Hunter's theme an unnerving, dissonant sense of movement.

For the planets, Harlin reviewed concept art and consulted the game's writers to get a sense of what each world was like, geologically as well as dramatically. On Belsavis, metallic percussion plays against organic sounds to suggest a world at odds with itself; hammered dulcimers and medieval harmonic structures lend Alderaan a sense of antiquity; solo tubas evoke Jabba's theme on Hutta. The composers referenced the Force theme for Voss, using glass instruments and bowed vibraphones for a shimmering, fragile sound that Harlin describes as "dust in a sunbeam. As if it were a soap bubble—any quick movements and suddenly the whole world around you would just pop and be gone."

The score would need to match the emotional content of each moment, changing on the fly depending on in-game events, so it was vital that cues blend into one another seamlessly. Harlin used an application called Wwise to facilitate transitions and to control cross-fades and where musical cues come in and out of one another. During game play, cues are placed into randomized "buckets," organized by mood, and they are pulled out as game-play events dictate. All cues were written without endings, so they fade into silence if no music is needed, allowing ambient environmental sounds to come smoothly to the forefront. The result: a pliable, adaptable soundtrack that responds to player choice, just as the game's characters and story-lines do. "We're trying to create material that isn't just musical wallpaper," says Harlin. "We want the music to function as a character in the world and be something that interacts with you and helps to augment your experience, giving it an emotional underpinning."

NOT JUST MUSICAL WALLPAPER

Life Is a Huttese Cabaret

Harlin recruited Peter McConnell (Grim Fandango, Brütal Legend) to compose ninety minutes of new cantina music: one-third consisting of quirky space jazz as played in the Mos Eisley cantina; another third making use of ARP synthesizers to create bizarre alien funk, reminiscent of the Max Rebo band in *Return of the Jedi*. For the last third, Harlin gave McConnell an open canvas. "One of the things we said was, 'Maybe this group of material is separate and heard only in Imperial cantinas?'" says Harlin. So what would Imperials listen to? "Peter decided that the Imperial army, being a fascist army, has a thing for cabaret music, much in the way the Nazis did in occupied France; so he wrote thirty minutes of Huttese cabaret music." McConnell worked from November 2009 through February 2010 writing, recording, and even providing both male and female vocals for the performances. "It's the most genius material I've heard," says Harlin.

ACKNOWLEDGMENTS

The authors would like to thank the following folks:

BioWare
Diego Almazan, Gabe Amatangelo, Cory Butler, Rob Chestney, Andrew Collins, Bill Dalton, Todd Davies, Ryan Dening, Dallas Dickinson, Jeff Dobson, Mark Howe, Arnie Jorgensen, Shawn Ketcherside, Emmanuel Lusinchi, Paul Marino, Jason Minor, Ray Muzyka, James Ohlen, Leo Olebe, Brad Prince, Damion Schubert, Deborah Shin, Rich Vogel, Gordon Walton, Clint Young, Greg Zeschuk, and Georg Zoeller

LucasArts
William Beckman, Mary Bihr, Hez Chorba, Rob Cowles, Roger Evoy, Matthew Filbrandt, Jesse Harlin, Orion Kellogg, Peter Kingsley, Jake Neri, Darragh O'Farrell, Tim Temmerman, and Derek Williams

Lucasfilm, Ltd.
Troy Alders, Leland Chee, Carol Roeder, Stacey Leong, Howard Roffman, and Sue Rostoni

Special thanks to Pablo Hidalgo and Steve Sansweet for much-needed feedback

Chronicle Books
Yolanda Cázares, Becca Cohen, Jake Gardner, Emily Haynes, Sarah Malarkey, and Emilie Sandoz

Public
Todd Foreman and Tessa Lee

Special thanks to Penny Arcade

And to George Lucas